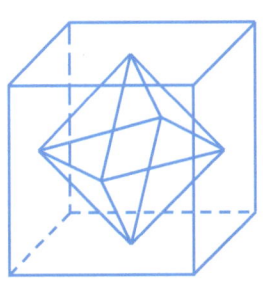 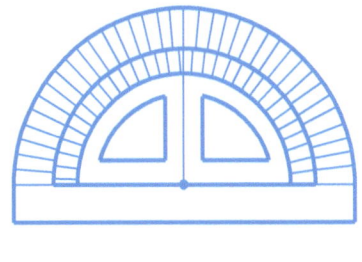 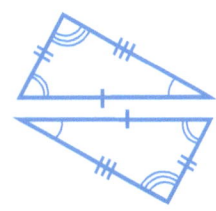 

A Basic Course
in Geometry

– Part 1 of 5

Bill Lembke

Citrus Software Publishing

Citrus Ridge, Florida

Published by Citrus Software Publishing, a Division of Citrus Software Corporation

Printed and bound in the United States of America.

A Basic Course in Geometry – Part 1 of 5

ISBN-13: 978-1477519578

ISBN-10: 1477519572

List of Topics

Chapter 9 – Two Dimensional Non-polytopes

Chapter 10- Three Dimensional Non-polytopes

Part 4 of 5

Chapter 11 – Spherical Geometry

Chapter 12 – Geometric Constructions

Chapter 13 – Geometric Proofs

- 13-1 Introduction
- 13-2 Why Learn Geometric Proofs?
- 13-3 Terms
- 13-4 Postulates and Theorems
- 13-5 How to Write a Proof
- 13-6 Types of Proofs
- 13-7 Summary
- Chapter Test

Chapter 14 – Assessment

- Chapters 1 – 13

Part 5 of 5

Appendix A – Glossary of Terms

Appendix B – Chapter Tests Solutions

Appendix C – Assessment Solutions

Index

List of Figures

Chapter 5 – Triangles and Quadrilaterals

List of Tables

Preface

Introduction

Geometry is a visual branch of mathematics. Almost every concept or problem is accompanied by a figure or table. The best way to learn something new is to actively participate, rather than be a passive observer. To successfully learn geometry, it is highly recommended that students try to create each drawing and solve each problem of this textbook.

"I hear and I forget. I see and I remember. I do and I understand." – Confucius, Chinese philosopher (551 BCE – 479 BCE)

The drawings you create should be made as accurately as possible, as each represents a geometric concept or step in solving a problem. If you are to learn by doing, you need to do it right. To create geometry drawings, the following tools are needed: (1) notebook paper – to record notes, theorems, problems, and solutions; (2) graph paper – to align and size drawn objects; (3) pencils – to write notes, concepts, and other recorded items; (4) colored pencils – to distinguish among parts of a drawing and drawing steps; (5) eraser – to correct errors and remove temporary lines; (6) ruler, with inch or centimeter markings – to measure distances and draw straight lines; (7) protractor, with angle markings – to draw and measure the angle sizes; and (8) compass – to draw circles and arcs.

To be successful in geometry, you should try to develop the following qualities into your learning style and philosophy. *Imagination*: You will need to visualize in your mind how to make a drawing or solve a problem and the how the various parts relate to each other. Only then should you actually start. Having a plan before you start will make the process quicker and more thoroughly complete. *Creativity*: You will need to determine the step-by-step process for solving a problem. There are often several possible approaches that may or may not be the best solution. *Curiosity*: You need to develop a desire to learn or know. Things that are novel or extraordinary should arouse intense interest. There are always new things to learn and to know why things are as they appear. *Intrepidness*: You need to develop outstanding courage and become fearless. Solving problems require bold and expressive steps. Sometimes you learn more from your mistakes, then always being correct. You should embrace challenges as opportunities to test your knowledge and abilities. The best motivation is self-motivation.

This textbook contains 461 figures and 150 tables to explain geometric concepts and examples. The subjects covered by this textbook include the following: (1) polytopes – objects drawn with straight lines, from polygon and polyhedron to polychoron and polyxennon; (2) non-polytopes – objects drawn with curved lines, from circle and ellipse to sphere and cylinder; (3) spherical geometry – geometry on a curved surface, rather than on a flat plane; (4) geometric constructions – drawings made using only a straightedge and compass; and (5) geometric proofs – step-by-step instructions used to prove theorems are valid.

Geometry is a fun type of mathematics. You will learn many new and interesting things during this geometry course. Are you ready to begin your educational journey? When you turn to the first chapter, your journey will begin.

Instructional Methods

The structure of this textbook and the instructional methods used will allow you to successfully learn geometry. A willing and motivated student can be taught any subject. This textbook uses the ABC Method of Instruction. It was developed by the Citrus Software Corporation, which has been creating computer software applications and instructional material since 1981. The ABC Method has since been used to teach a wide variety of academic subjects and has been used successfully by students for more than 30 years.

The ABC Method motto is "We make the complex simple." It starts with the premise that all complex subjects can be made simple, if broken down into their basic parts. After you learn the basic parts of the subject, the parts are combined together to make more complex parts. The complex subject has been made simple.

The ABC Method gets its name from the first three letters of the alphabet. When you are learning a language, you start by learning the letters. The letters are combined to make words, and the words then become sentences. This leads to the documents and conversations of effective communication. The complex exchange of ideas begins with the simple letters of the alphabet.

Geometry includes many mathematical topics such as angles, polytopes, spherical geometry, and geometric proofs. This may seem like a complex task, but it will be divided into many small tasks. This will be an ideal use of the ABC Method of Instruction.

About the Author

Bill Lembke has more than 30 years of experience as a computer engineer, business manager, and instructor. He began his career as a computer programmer and analyst designing commercial software applications, and then later worked in a similar capacity for federal government agencies. After twenty years as a computer engineer and project manager, he changed careers to become a full time college professor. Immediately prior to writing this book, he was a professor of computer science and mathematics for 10 years. As department chairperson and assistant director of education, he established curriculum standards and developed course content for several degree programs. He is currently conducting academic research and developing mathematical course curriculum. Bill Lembke has a Masters of Education degree in Instructional Technology, along with other advanced degrees, and has been a guest lecturer at several colleges and universities. He has also served as an educational consultant to schools, corporations, government agencies, and other organizations. He has written thousands of articles and many books on a wide variety of academic subjects.

The Book

This textbook is a basic course in geometry. It assumes the student has little or limited knowledge of geometry, which means terms and concepts are explained before they are extensively used. It starts with basic concepts, and then builds upon them to develop more complex ideas. This book is dedicated to everyone with an interest in geometry, wanting clear and concise definitions and examples of basic to complex concepts.

The material presented in this textbook can be used in either a standards-based curriculum or a traditional-based curriculum. The individual instructor or school administration will determine which curriculum will be used in the geometry course. Instructors can supplement the text with additional problems and examples as necessary to match their instructional methods and learning environment. Extensive research has been made to ensure proper definitions and term usages.

The content of the book has been used and tested in a variety of grade levels for effectiveness, and developed over a period of several years. The appropriate grade level of this textbook will be determined by the educators and administrators. It may range from advanced middle school to early college, depending upon which chapters are used and the amount of supplemental material included in the course. Some students may also use this textbook as a review of geometry before studying more advanced mathematical subjects.

Book Content

Each of the chapters, 1-13, explains a group of related geometric topics with detailed descriptions and examples. There are 13 chapter tests. Chapter 14 is the comprehensive final exam. Appendixes and an index follow Chapter 14. The chapters are summarized below.

Chapter 1 – Concepts and Standards is divided into two sections. The first section, Geometry Concepts, describes the history of geometry and describes basic geometric concepts. The second section, Geometry Standards, provides recommendations and guidelines for curriculum improvement.

Chapter 2 – Angles describes angle rotation, labeling, and measurement. Angle types and comparisons are explained.

Chapter 3 – Polytopes describes hyperplanes and dimensions. Polytope types and decomposition are explained.

Chapter 4 – Polygons describes polygon types, parts, and properties. Formulas for regular polygons and triangles are explained.

Chapter 5 – Triangles and Quadrilaterals is divided into two sections. The first section, Triangles, describes triangle types, parts, and congruency. Points, lines, and circles of a triangle and formulas are explained. The second section, Quadrilaterals, describes and explains quadrilateral types, properties, and formulas.

Chapter 6 – Polyhedron describes polyhedron types and naming conventions. Polyhedron symmetry, characteristics, and nets are explained.

Chapter 7 – Polyhedron Solids – Part 1 describes Platonic, Kepler-Poinsot, Archimedean, Catalin, and Johnson solids. Formulas and examples are explained.

Chapter 8 – Polyhedron Solids – Part 2 describes pyramids, bipyramids, trapezohedron, frustum, prism, antiprism, wedge, toroidal polyhedron, and compound polyhedron. Formulas and examples are explained.

Chapter 9 – Two Dimensional Non-polytopes describes conic sections, circle, ellipse, parabola, hyperbola, circular sector, circular segment, circular ring, circular ring sector, stadium, and oval.

Chapter 10 – Three Dimensional Non-polytopes describes sphere, cone, cylinder, conical frustum, torus, spherical cap, spherical sector, spherical segment, spherical wedge, and capsule.

Chapter 11 – Spherical Geometry compares plane and spherical geometry. Spherical geometry properties, polygons, and tessellations are described. Spherical angles, measurements, triangle congruency, and formulas are explained.

Chapter 12 – Geometric Constructions describes geometry tools, construction types, and construction rules. 14 construction problems with step-by-step solutions are provided.

Chapter 13 – Geometric Proofs describes proofs, proof types, and proof terms. 7 geometric proof problems with step-by-step solutions are provided.

The Cover

Cover design: Sunrise – Each day brings opportunities to learn something new.

Let today be the beginning of your journey on your path to enlightenment and self-actualization.

- ## 1-1 Introduction

Geometry (Ancient Greek: γεωμετρία; *geo*-"earth", *-metri* "measurement") is a branch of mathematics which studies spatial relationships and spatial structures. It is concerned with the properties and relationships of points, lines, angles, curves, surfaces, and solids. Geometric principles have evolved over the millennia from human observations and interactions with the physical world. Many ancient civilizations in the Middle East (Egyptian and Babylonian), Central and Eastern Asia (Hindu and Chinese), and Central America (Olmec and Mayan) developed collections of measurements and procedures. Comparisons of physical shapes and sizes led to calculations of distance, development of numbering systems, construction techniques, and astronomical observations.

The geometric knowledge of ancient civilizations was often developed over long periods of time and may have included the results of experimentation, casual observation, estimations, guessing, and superstitions. The results were often approximate values sufficient for practical purposes. A codification system for geometry began with the ancient Greeks. The Greeks were first to take the step from known properties of geometric figures to a system of logic, which could be used to derive unknown properties.

Geometry was thoroughly organized in about 300 BCE, when the Greek mathematician Euclid gathered what was known at the time, added original work of his own, and arranged 465 propositions into 13 books, called '*Elements*'. The books covered not only plane and solid geometry but also much of what is now known as algebra, trigonometry, and advanced arithmetic. His book, *Elements*, is widely considered the most influential textbook of all time, and was known to all educated people in the West until the middle of the 20th century.

Geometry was revolutionized by Euclid, who introduced mathematical rigor and the axiomatic method still in use today. Euclid founded his geometry on a small group of definitions, postulates, and axioms, which he accepted as self-evident. Step by step from these he deduced the whole set of theorems of classical geometry. His achievements still stand, and much of it is included in the elementary geometry taught today. It involves plane geometry, which treats figures on a flat scale, and solid geometry, which treats figures in three dimensions. This branch of mathematics is often called Euclidean Geometry in his honor.

The next step in geometry was the development of solid geometry. In mathematics, solid geometry is the traditional name for the geometry of three-dimensional space — for practical purposes the kind of space we live in. It was developed following the development of plane geometry, which requires all points to be on a plane. ==Stereometry== deals with the measurements of volumes and dimensions of various solid figures including cylinder, circular cone, truncated cone, sphere, and prism. It should be noted that trigonometry, which is usually taught as a separate branch of mathematics, is subset of geometric principles concerned with the measurement of triangles. Some trigonometric terms will be discussed in the overall topic of geometry.

Discovery Activity.

1. Discover the history of geometry. Select one ancient civilization from each of the three geographic areas (Middle East, Central and Eastern Asia, and Central America). Conduct research to explore the numbering systems and geometric principles of each selected civilization.
   a. Make a table to show the numbering systems and geometric principles of each of the selected civilization.
   b. Make a table to show the similarities and differences among each of the selected civilizations.
   c. Make a table that compares the information of the selected civilizations with the numbering system and geometric principles of the ancient Greeks.
2. Read Euclid's *Elements*. It is available in print and online from many sources.
   a. Make a list of the definitions, postulates, and common notations (axioms).
   b. Make a list and summarize the ten most important of the 465 propositions.
   c. Describe the influence that the *Elements* had on the development of geometry.

• 1-2 The Geometry Universe

There are three worlds in the geometry universe. The first world is called plane geometry and it exists in two dimensions. The second world is called solid geometry and it exists in three dimensions. The third world is called spherical geometry and it exists in two dimensional on the surface of a sphere. The world of plane geometry is viewed from above, with the x-axis appearing as east and west and the y-axis appearing as north and south. The place where the x-axis and y-axis intersect is called the origin. On the plane, which is a flat surface extending infinitely in all directions, you can draw points, lines, and two-dimensional objects.

Measurements of the two-dimensional objects can be made, such as the length of the sides, the angles between the lines, and the area enclosed by the lines. However, nothing can ever rise above or sink below the surface of the plane.

The world of ==plane geometry== involves zero, one, and two dimensional objects. The plane is a flat surface, endless in all directions. It contains an unlimited number of points. The plane has no thickness and must be viewed from above. In the two-dimensional world of plane geometry, the plane is the universe of all points. A point can be located anywhere on the surface of the plane. A point is zero-dimensional. It is a location, without size, width, height, depth, area, or volume. A point is given a visual representation as a dot. If a series of points are connected together, side by side in a straight queue, they become a line. A line has length, extending infinitely in two opposite directions, and is one-dimensional. A line with an end point and extending infinitely only in one direction is called a ray. A line with two end points, extending only between the two end points, is called a line segment. Line segments can be connected together at their end points to create shapes, such as triangles and squares. These shapes, as a group, are called polygons. Polygons are two-dimensional. The shape and size of a polygon can be measured and compared to other polygons. This is the essence of plane geometry.

The world of ==solid geometry== is viewed from above or from any side. The x-axis and y-axis are the same as in the plane geometry world, but the z-axis is added to allow for objects to rise above or sink below the plane. There can be several planes, which can be parallel or intersect at various angles. Measurements of the three-dimensional objects include everything of the plane geometry world, in addition to volume. Volume is the space contained within a three-dimensional object. Higher levels of dimension involve changing three-dimensional objects by duplication and other transformations. (Caution: The higher levels of geometry dimensions should not be confused with the higher levels of physical dimensions used in advanced physics to describe time and space, astronomy, and subatomic particles.) The world of solid geometry involves three dimensional objects. The points of solid geometry can be on the plane, or above or below the plane. In a world of three-dimensions, a point can be located anywhere. Objects can contain zero, one, two, or three dimensions. A three-dimensional polyhedron is created by combining together two-dimensional polygons. The possible shapes and objects are unlimited. The objects are unbounded by any constraints. The universe is all points anywhere. This is the essence of solid geometry.

The world of ==spherical geometry== is viewed from above. It is the study of two-dimensional figures on the surface of a sphere. A sphere is a perfectly round geometrical object in three-dimensional space. Like a circle in two dimensions, a perfect sphere is completely symmetrical around its center, with all points on the surface lying the same distance from the center point, the radius. The curve formed by a plane cutting a sphere is a circle. If the plane goes through the center of the sphere, the circle is called a great circle of the sphere. It is the largest circle that can be drawn upon the sphere, and all great circles of the same or equal spheres are of equal size. The shortest distance between two points on a spherical surface, measured on the surface, is the distance along the great circle through those points. A plane cutting a sphere in a great circle divides the sphere into two equal segments called hemispheres. The diameter of a sphere is the diameter of one of its great circles. In spherical geometry, the rules of plane and solid geometry are modified. Parallel lines and straight lines do not exist. However, two sided polygons, called biangles, do exist. Triangles have more than 180 degrees and a triangle with three right angles is possible. The angles of a quadrilateral are greater than 90 degrees. A square and a pentagon can both have 120 degree angles. In plane geometry, the shortest distance between two points is called a straight line. In spherical geometry, the shortest distance is a curved line called a geodesic. The universe is all points on the surface of a sphere. This is the essence of spherical geometry.

1-3 Definitions

A definition of basic terms is needed to begin a discussion of geometry.

Point – A point is the simplest object in geometry. A point is a location. It does not have a size, width, height, depth, area, or volume. A dot is used as a visual representation of a point. A point can be at any location, such as on a plane or on another object. If an object is magnified to show greater detail, any points on the object do not increase in size.

Line – A line is a series of points extending infinitely in a straight path in two opposite directions. It is the representation of the location of points. A line has length or width, but does not have height, depth, area, or volume.

Line Segment – A line segment is part of a line, with points marking the beginning and ending locations.

Ray – A ray is half of a line. It has a point marking the beginning location and extends infinitely in one direction.

Angle – An angle is the space between two rays or line segments that meet at a common end point, called the vertex. An angle can be measured in degrees or radians. Two intersecting lines, rays, or line segments create four adjacent angles.

Plane – A plane is a flat surface extending infinitely in all directions. It is made up of points and does not have a thickness. Objects can be placed on a plane to improve the visualization of properties and location relationships. In plane or two-dimensional geometry, the universe is all points on the plane.

Width – Width is the horizontal distance on a plane. In coordinate systems, it is labeled the X axis.

Height – Height is the vertical distance on a plane. In coordinate systems, it is labeled the Y axis.

Depth – Depth is the volume distance of an object on an intersecting perpendicular plane. In a coordinate system, it is labeled the Z axis. In solid or three-dimensional geometry, the universe is all points in space.

Dimensions – Dimensions are used to describe objects in physical space. The dimension of an object is defined as the minimum number of coordinates needed to specify each point within it. A point has zero dimensions. A line has one dimension. A polygon has two dimensions. A polyhedron has three dimensions.

Polygon – A polygon is a two-dimensional object bounded by straight lines, such as a trigon (triangle), tetragon (quadrilateral), or pentagon.

Polyhedron – A polyhedron is a three-dimensional object bounded by polygons, such a pyramid, cube, or prism.

Mathematical pun: Without geometry, life is pointless.

• 1-4 Geometry Objects

Geometry is the mathematics of shapes and space. Various shapes can be identified, measured, and compared. Visualization is an important part of geometrical thinking. Creating a physical model of each object is too time-consuming. Making a two-dimensional drawing to express a three-dimensional shape requires imagination. The process of observing and identifying geometric objects includes five steps. The five object identification steps are shown in the table below. A sample object, next to the table, will be identified using these steps.

Object Identification Steps
1. Polytope or nonpolytope
2. Two or three dimensions
3. Measurement
4. Comparison
5. Summarize

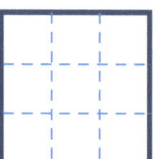

Table 1-1: Identification Steps   Figure 1-1: Sample Object

The first step is to determine if an object is made of straight lines or curved lines. There are two groups of objects in geometry, polytopes and nonpolytopes. Polytopes are objects created from straight lines. A line is assumed to be straight, unless otherwise stated. A straight line minimizes the distance between points, on a flat surface such as a plane. A curved line is a series of points not in a straight queue and can change direction one or several times. On a curved surface, such as a sphere, the shortest distance between points is a curved line. Objects that contain curved lines are nonpolytopes. The sample object is a polytope.

The second step is to determine the number of dimensions contained in the object. Both polytopes and nonpolytopes are divided into groups based on the number of dimensions they contain. A two-dimensional polytope is called a polygon. Polygons include triangle, square, and pentagon. A three-dimensional polytope is called a polyhedron. Polyhedra include tetrahedron, cube, and octahedron. Two-dimensional nonpolytopes include circle, ellipse, and ring. Three-dimensional nonpolytopes include sphere, cylinder, and cone. In the first figure below, two-dimensional and three-dimensional polytopes are shown. In the second figure below, two-dimensional and three-dimensional nonpolytopes are shown. The sample object has two dimensions and has the shape of a square.

Graphs of two-dimensional polytopes

Figure 1-2:	Figure 1-3:	Figure 1-4:
Triangle	Square	Pentagon
(3 sides)	(4 sides)	(5 sides)

Graphs of three-dimensional polytopes

Figure 1-5:	Figure 1-6:	Figure 1-7:
Tetrahedron	Cube	Octahedron
(4 faces)	(6 faces)	(8 faces)

Graphs of two-dimensional nonpolytopes

| Figure 1-8: | Figure 1-9: | Figure 1-10: |
| Circle | Ellipse | Ring |

Graphs of three-dimensional nonpolytopes

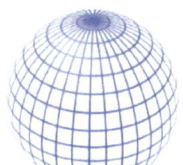  

| Figure 1-11: | Figure 1-12: | Figure 1-13: |
| Sphere | Cylinder | Cone |

The third step is to measure the object so it can be more specifically identified. The shape of an object is its overall appearance. If the shape has straight sides, how many sides does it have? If the shape has curved sides, is it round like a ball or oblong? The size of the object is a comparison of the object relative to the size of other objects. More specific measurements can be made using formulas. In the figure below, the shapes square, circle, cube, and sphere are shown. The sample object can be measured as a square.

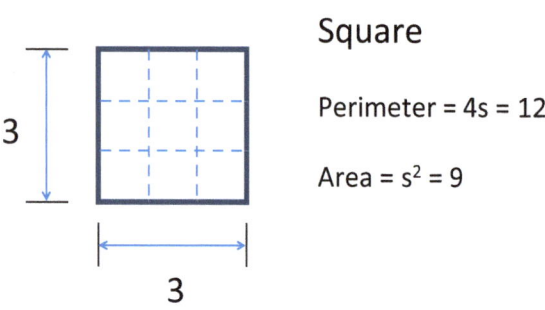

Square

Perimeter = 4s = 12

Area = s^2 = 9

Figure 1-14: Square Measurement

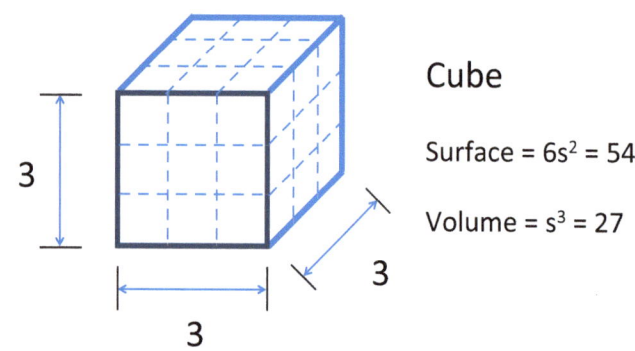

Cube

Surface = $6s^2$ = 54

Volume = s^3 = 27

Figure 1-15: Cube Measurement

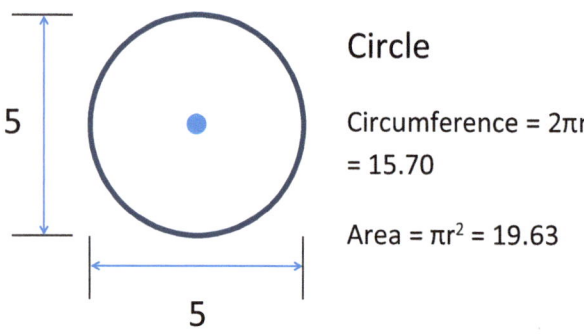

Circle

Circumference = 2πr
= 15.70

Area = $πr^2$ = 19.63

Figure 1-16: Circle Measurement

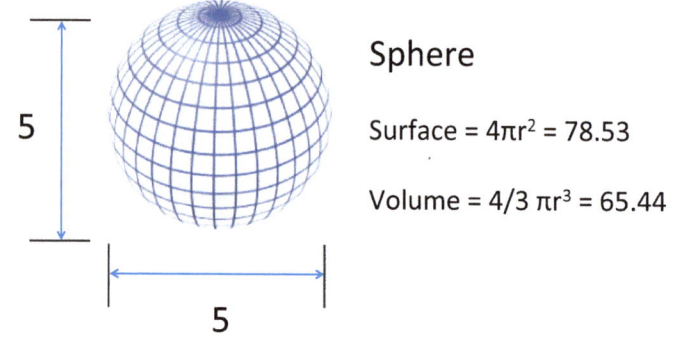

Sphere

Surface = $4πr^2$ = 78.53

Volume = 4/3 $πr^3$ = 65.44

Figure 1-17: Sphere Measurement

The fourth step is to compare the object to other objects. Four sided shapes are called quadrilaterals. Using the figure above, the sample object matches the first shape and is a square. What makes it a square? The table below shows the seven types of quadrilaterals based on special properties.

Quadrilateral Name	Properties
Kite	2 pairs of adjacent equal sides
Trapezoid	1 pair of parallel sides
Isosceles Trapezoid	1 pair of parallel sides and 1 pair of equal sides
Parallelogram	2 pairs of parallel sides
Rhombus	2 pairs of parallel sides and 4 equal sides
Rectangle	2 pairs of parallel sides and all angles equal to 90 degree angles
Square	2 pairs of parallel sides, all angles equal to 90 degrees, and 4 equal sides

Table 1-2: Quadrilateral types

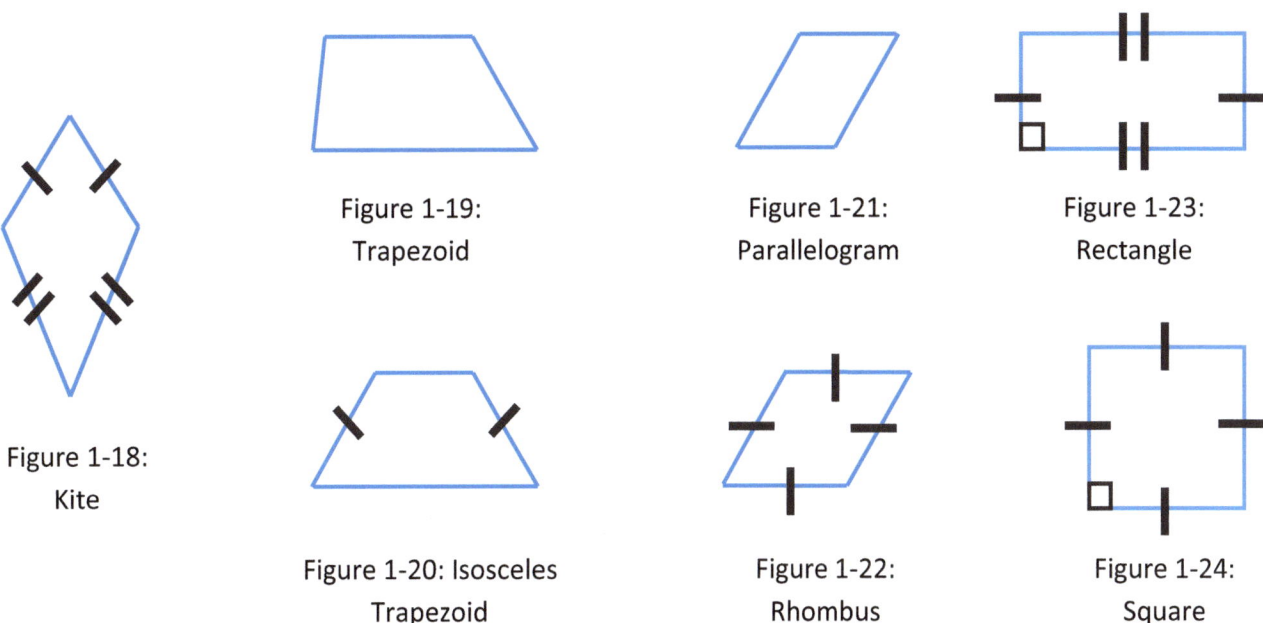

Figure 1-19:
Trapezoid

Figure 1-21:
Parallelogram

Figure 1-23:
Rectangle

Figure 1-18:
Kite

Figure 1-20: Isosceles
Trapezoid

Figure 1-22:
Rhombus

Figure 1-24:
Square

In the figure above, the seven types of quadrilaterals are shown. The quadrilaterals share properties, such as parallel sides or equal sides. Four of the shapes have 2 pairs of parallel sides, including the square. Two of the shapes have all angles equal to 90 degrees, including the square. Two of the shapes have 4 equal sides, including the square. A combination of the three previous statements uniquely identifies a square.

The fifth step is summarizing all of the properties of the object. The sample object has been determined to be a square. The properties of the object can be summarized. It is a polytope with two dimensions. By identifying the shape correctly as a square, formulas can be used to calculate the measurements of the object. The size is 3 units by 3 units. The perimeter, or measure around the outside, is 12 units. The area, or amount of units within the object, is 9 square units. The selection of the object as being a square can be verified as being correct by using the table of properties for quadrilaterals. The drawing of a square allows for a matching of the shape to an established standard for a square.

After completing all five steps, the object has been fully described as a unique object. Using a systematic approach will ensure a complete and accurate analysis of the object.

• 1-6 Summary

Geometry is a branch of mathematics which studies spatial relationships and spatial structures. It is concerned with the properties and relationships of points, lines, angles, curves, surfaces, and solids. A codification system for geometry began with the ancient Greeks. The Greeks were first to take the step from known properties of geometric figures to a system of logic, which could be used to derive unknown properties.

There are three worlds in the geometry universe. The first world is called plane geometry and it exists in two dimensions. The second world is called solid geometry and it exists in three dimensions. The third world is called spherical geometry and it exists in two dimensional on the surface of a sphere.

A definition of basic terms is needed to begin a discussion of geometry. A point is the simplest object in geometry. A point is a location. A line is a series of points extending infinitely in a straight path in two opposite directions. A line segment is part of a line, with points marking the beginning and ending locations. A ray is half of a line. An angle is the space between two rays or line segments that meet at a common end point, called the vertex. A plane is a flat surface extending infinitely in all directions. Width is the horizontal distance on a plane. Height is the vertical distance on a plane. Depth is the volume distance of an object on an intersecting perpendicular plane. The dimension of an object is defined as the minimum number of coordinates needed to specify each point within it. A polygon is a two-dimensional object bounded by straight lines.

Geometry is the mathematics of shapes and space. Various shapes can be identified, measured, and compared. The process of observing and identifying geometric objects includes five steps. The first step is to determine if an object is made of straight lines or curved lines. The second step is to determine the number of dimensions contained in the object. The third step is to measure the object so it can be more specifically identified. The fourth step is to compare the object to other objects. The fifth step is summarizing all of the properties of the object. After completing all five steps, the object has been fully described as a unique object. Using a systematic approach will ensure a complete and accurate analysis of the object.

● **1-7 Understanding Geometry**

A thorough understanding of geometry, unlike most branches of mathematics, can not be accomplished by mere rote memorization. The solving of geometric problems often requires inductive and deductive reasoning and problem solving skills. Students cannot be expected to prove geometric theorems until they have built up an extensive understanding of the systems of relationships between geometric ideas. These systems can not be learned by rote, but must be developed through familiarity by experiencing numerous examples and counterexamples, the various properties of geometric figures, the relationships between the properties, and how these properties are ordered.

The National Council of Teachers of Mathematics (NCTM) has published several documents with recommendations and guidelines that provide a framework and set of goals for mathematics curriculum improvement. Among these publications are the *Curriculum and Evaluation Standards* (1989) and the *Principles and Standards for School Mathematics* (2000). Standards-based curricula have been developed to try to meet the recommendations made by mathematics education researchers and the needs of mathematics students. The intent of the *Standards* documents (1989, 2000) and standards-based curricula is that students will explore, conjecture, reason logically, and use several different methods to solve problems. NCTM *Standards* documents (1989, 2000) address issues regarding the teaching and learning of geometry. The *Standards* (1989) states: Evidence suggests that the development of geometric ideas progresses through a hierarchy of levels. Students first learn to recognize whole shapes and then to analyze the relevant properties of shapes. Later they can see relationships between shapes and make simple deductions. Curriculum development and instruction must consider this hierarchy.

The National Council of Teachers of Mathematics (NCTM) published *Principles and Standards for School Mathematics* in 2000 with specific geometry standards for grades 9-12. The NCTM has a standard specific to Geometry. The following table shows the geometry standard.

Standards: Instructional programs from prekindergarten through grade 12 should enable all students to—	Expectations: In grades 9–12 all students should—
Analyze characteristics and properties of two- and three-dimensional geometric shapes and develop mathematical arguments about geometric relationships	• analyze properties and determine attributes of two- and three-dimensional objects; • explore relationships (including congruence and similarity) among

	classes of two- and three-dimensional geometric objects, make and test conjectures about them, and solve problems involving them; • establish the validity of geometric conjectures using deduction, prove theorems, and critique arguments made by others; • use trigonometric relationships to determine lengths and angle measures.
Specify locations and describe spatial relationships using coordinate geometry and other representational systems	• use Cartesian coordinates and other coordinate systems, such as navigational, polar, or spherical systems, to analyze geometric situations; • investigate conjectures and solve problems involving two- and three-dimensional objects represented with Cartesian coordinates.
Apply transformations and use symmetry to analyze mathematical situations	• understand and represent translations, reflections, rotations, and dilations of objects in the plane by using sketches, coordinates, vectors, function notation, and matrices; • use various representations to help understand the effects of simple transformations and their compositions.
Use visualization, spatial reasoning, and geometric modeling to solve problems	• draw and construct representations of two- and three-dimensional geometric objects using a variety of tools; • visualize three-dimensional objects and spaces from different perspectives and analyze their cross sections; • use vertex-edge graphs to model and solve problems; • use geometric models to gain insights into, and answer questions in, other areas of mathematics; • use geometric ideas to solve problems in, and gain insights into, other disciplines and other areas of interest such as art and architecture.

Table 1-3: Geometry Standards

One of the expectations of standard one is to "establish the validity of geometric conjectures using deduction, prove theorems, and critique arguments made by others." This shows that higher order thinking is expected of high school students. One of the expectations of standard four is to "draw and construct representations of two- and three-dimensional geometric objects using a variety of tools." This expectation shows the need for geometric constructions using a variety of tools such as pencil, paper, compass, and straightedge, along with geometry software.

The NCTM has a reasoning and proof standard for grades 9-12. The following table shows the reasoning and proof standard.

Instructional programs from prekindergarten through grade 12 should enable all students to—
<ul><li>recognize reasoning and proof as fundamental aspects of mathematics;</li><li>make and investigate mathematical conjectures;</li><li>develop and evaluate mathematical arguments and proofs;</li><li>select and use various types of reasoning and methods of proof.</li></ul>

Table 1-4: Reasoning and Proof Standards

This standard shows the need for geometric proofs. Geometry software, such as *The Geometer's Sketchpad* ®, can be used to explore and make conjectures, but it does not prove a mathematical statement. Proofs formally present the results of mathematical thought using logic and deductive reasoning as a method for establishing the validity of ideas. Learning geometric proofs will develop analytical thinking skills. The skills needed to develop geometric proofs will enable the student to be successful in other areas of mathematics and subjects.

Discovery Activity.

1. Discover geometry standards. Select Common Core State Standards Initiative and read the Common Core State Standards for Mathematics and Mathematics Appendix A (http://www.corestandards.org/the-standards). Select another state or regional department of education. Conduct research to explore the geometry standards for grades 9-12.
    a. Make a table to show the standards for instruction and the expectations for students of the Common Core State Standards and the other selected department of education.

     b. Make a table to show the similarities and differences between the Common Core State Standards and the other selected department of education.

     c. Make a table that compares the information of the Common Core State Standards and the other selected department of education with the standards of the National Council of Teachers of Mathematics (NCTM).

2. Design geometry standards for grades 9-12. Review several geometry textbooks.

     a. Make a list of topics that should be included in a course in geometry.

     b. Make a lesson plan for a course in geometry. In what order should the topics be learned? How much time should be spent on each topic?

     c. Make a list of classroom activities that would be both fun and informative, to make learning geometry easier.

Other department of education standards, not part of the Common Core State Standards:

Texas State Standards (§111.34.) – Geometry:
http://ritter.tea.state.tx.us/rules/tac/chapter111/ch111c.html

Virginia State Standards – Geometry:
http://www.doe.virginia.gov/testing/sol/standards_docs/mathematics/index.shtml

Ontario Providence Standards – Measurement and Geometry:
http://www.edu.gov.on.ca/eng/curriculum/secondary/mathtr9curr.pdf

1-8 Van Hiele Levels of Geometric Understanding

A widely used model for geometry instruction and student learning was developed by a husband-and-wife team of Dutch educators, Pierre van Hiele and Dina van Hiele-Geldof (1959). They noticed the difficulties that their students had in learning geometry. These observations led them to develop a theory involving levels of thinking in geometry that students pass through as they progress from merely recognizing a figure to being able to write a formal geometric proof. Their theory explains why many students encounter difficulties in their geometry course, especially with formal proofs. The van Hieles believed that writing proofs requires thinking at a comparatively high level, and that many students need to have more experiences in thinking at lower levels before learning formal geometric concepts.

The following table shows the Van Hiele Levels of Geometric Understanding.

Level	Description
0	*Visualization*: Students recognize figures by appearance alone, often by comparing them to a known prototype. The properties of a figure are not perceived. At this level, students make decisions based on perception, not reasoning.
1	*Analysis*: Students see figures as collections of properties. They can recognize and name properties of geometric figures, but they do not see relationships between these properties. When describing an object, a student operating at this level might list all the properties the student knows, but not discern which properties are necessary and which are sufficient to describe the object.
2	*Abstraction*: Students perceive relationships between properties and between figures. At this level, students can create meaningful definitions and give informal arguments to justify their reasoning. Logical implications and class inclusions, such as squares being a type of rectangle, are understood. The role and significance of formal deduction, however, is not understood.
3	*Deduction*: Students can construct proofs, understand the role of axioms and definitions, and know the meaning of necessary and sufficient conditions. At this level, students should be able to construct proofs such as those typically found in a high school geometry class.
4	*Rigor*: Students at this level understand the formal aspects of deduction, such as establishing and comparing mathematical systems. Students at this level can understand the use of indirect proof and proof by contrapositive, and can understand non-Euclidean systems.

Table 1-5: Levels of Geometric Understanding

Most students begin geometry at level 0. Some students may find moving into level 1 challenging, particularly as they learn new terminology. At level 2, the students must learn to use the language of mathematics to express mathematical ideas precisely. As individual students investigate and present their ideas, their levels can be monitored. A more formal assessment is completed at the end of each chapter. The goal is to help all students move up to level 2 or higher by the end of the course.

• 1-9 Progress Indicators

Progress indicators allow for the assessment of students, to classify them by level. The following table shows progress indicators as the students' progress from level 0.

Level	Indicator
1	• Use of imprecise properties (qualities) to compare drawings and to identify, characterize, and sort shapes.

		• References to visual prototypes to characterize shapes.
		• Inclusion of irrelevant attributes when identifying and describing shapes, such as orientation of the figure on the page.
		• Inability to conceive of an infinite variety of shapes.
		• Inconsistent sortings; that is, sortings by properties not shared by the sorted shapes.
		• Inability to use properties as necessary conditions to determine a shape; for example, guessing the shape in the mystery shape task after far too few clues, as if the clues triggered a visual image.
2		• Comparing shapes explicitly by means of properties of their components.
		• Prohibiting class inclusions among several general types of shapes, such as quadrilaterals.
		• Sorting by single attributes, such as properties of sides, while neglecting angles, symmetry and so forth.
		• Application of a litany of necessary properties instead of determining sufficient properties when identifying shapes, explaining identifications, and deciding on a mystery shape.
		• Descriptions of types of shapes by explicit use of their properties, rather than by type names, even if known.
		• Explicit rejection of textbook definitions of shapes in favor of personal characterizations.
		• Treating geometry as physics when testing the validity of a proposition; for example relying on a variety of drawings and making observations about them.
		• Explicit lack of understanding of mathematical proof.
3		• Formation of complete definitions of types of shapes.
		• Ability to modify definitions and immediately accept and use definitions of new concepts.
		• Explicit references to definitions.
		• Ability to accept equivalent forms of definitions.
		• Acceptance of logical partial ordering among types of shapes, including attributes.
		• Ability to sort shapes according to a variety of mathematically precise attributes.
		• Explicit use of "if, then" statements.
		• Ability to form correct informal deductive arguments, implicitly using such logical forms as the chain rule (if p implies q and q implies r, then p implies r) and the law of detachment (modus ponens: if p, then q).
		• Confusion between the roles of axiom and theorem.
4		• Clarifications of ambiguous questions and rephrasing of problem tasks into precise language.
		• Frequent conjecturing and attempts to verify conjectures deductively.
		• Reliance on proof as the final authority in deciding the truth of a mathematical proposition.
		• Understanding the roles of the components in a mathematical discourse, such as axioms, definitions, theorems, and proof.
		• Implicit acceptance of the postulates of Euclidean geometry.

Table 1-6: Progress Indicator Levels

Three major distinctions among students within level 1 can be identified:

1. Infinite Variety: Students have a clear understanding of an infinite variety of shapes versus students who cannot verbalize an infinite variety of shapes, but their understanding of an infinite variety of shapes is near.
2. Necessary Properties: Necessary properties of shapes are clearly understood versus necessary properties of specific shapes are still being formulated.
3. Precise Language: Students use precise language to discuss or describe the components and properties of shapes versus students use imprecise, sometimes ambiguous, visual descriptions to discuss components of shapes.

The three distinctions described above divide students initially assessed at van Hiele level 1 into three subsets within level 1: level 1A, level 1B, and level 1C. Students are divided into these subsets according to their understanding of these three distinctions. For example, students within *level 1C* understand an infinite variety of geometric shapes, understand necessary properties of geometric shapes, and use precise language to discuss geometric concepts. By contrasts, students within *level 1A* cannot verbalize an infinite variety of geometric shapes, necessary properties of geometric shapes are still being formulated within the student's understanding, and they tend to use descriptive language rather than precise language to discuss geometric concepts.

1-10 Learning Phases

According to the van Hieles, a student progresses through each level of thought as a result of instruction that is organized into five phases of learning. The phases of learning are described in the table below.

Phase	Description
1	*Information or inquiry*: Students get acquainted with the material and begin to discover its structure. Teachers present a new idea and allow the students to work with the new concept. By having students experience the structure of the new concept in a similar way, they can have meaningful conversations about it. (A teacher might say, "This is a rhombus. Construct some more rhombi on your paper.")
2	*Guided or directed orientation*: Students do tasks that enable them to explore implicit relationships. Teachers propose activities of a fairly guided nature that allow students to become familiar with the properties of the new concept which the teacher desires them to learn. (A teacher might ask, "What happens when you cut out and fold the rhombus along a diagonal? What happens when you fold along the other diagonal?" and so on, followed by discussion.)
3	*Explicitation*: Students express what they have discovered and vocabulary is introduced. The students' experiences are linked to shared linguistic symbols. The van Hieles believe it is more profitable to learn vocabulary *after* students have had an opportunity to become familiar with the concept. The discoveries are made as explicit as possible. (A teacher might say, "Here are the properties we have noticed and some

		associated vocabulary for the things you discovered. Let's discuss what these mean.")
4		*Free orientation*: Students do more complex tasks enabling them to master the network of relationships in the material. They know the properties being studied, but need to develop fluency in navigating the network of relationships in various situations. This type of activity is much more open-ended than the guided orientation. These tasks will not have set procedures for solving them. Problems may be more complex and require more free exploration to find solutions. (A teacher might say, "How could you construct a rhombus given only two of its sides?" and other problems for which students have not learned a fixed procedure.)
5		*Integration*: Students summarize what they have learned and commit it to memory. The teacher may give the students an overview of everything they have learned. It is important that the teacher not present any new material during this phase, but only a summary of what has already been learned. The teacher might also give an assignment to remember the principles and vocabulary learned for future work, possibly through further exercises. (A teacher might say, "Here is a summary of what we have learned. Write this in your notebook and do these exercises for homework.") Supporters of the van Hiele model point out that traditional instruction often involves only this last phase, which explains why students do not master the material.

Table 1-7: Learning Phases

● 1-11 Connected Mathematics Project

The Connected Mathematics Project (CPM) reform-based instructional model, "launch, explore, and summarize", is a problem-centered teaching approach based on the NCTM standards that opens the classroom to exploring, conjecturing, reasoning, and communicating. Connected Mathematics, developed by Michigan State University, addresses both the content and the process standards of the NCTM. The curriculum is divided into units, each of which contains investigations with major problems that the teacher and students explore in class. Extensive problem sets are included for each investigation to help students practice, apply, connect, and extend these understandings. The process standards are: Problem Solving, Reasoning and Proof, Communication, Connections, and Representation.The model of instruction involves three main phases, the *launch* phase, the *explore* phase, and the *summarize* phase.

In the *launch* phase, the teacher introduces the problem to the class. The teacher makes sure the students understand the problem and are engaged in it. It is important that the problem be interesting and makes connections with earlier concepts in mathematics or with past experiences of students. It is also an opportunity for the teacher to introduce a new idea.

In the *explore* phase, students work on the task individually, in small groups, or occasionally as a whole class. The students work on the problem by searching for patterns, gathering data, trying special cases, making conjectures, and exchanging ideas. The teacher moves around the classroom, observing student's work and offering help as appropriate. The teacher provides encouragement and confirmation to on-track behavior, and may ask guiding questions and offer redirections when needed. He or she may also offer additional challenges for those students who quickly solve the problem or may not find it challenging enough.

The *summarize* phase begins when most students make sufficient progress towards a solution to the problem. Students discuss their solutions and share the strategies they used to reach a solution. They will appreciate other approaches to the problem, and can see ways to enhance their own strategies. The teacher also offers guidance and suggestions for a deeper understanding of the concepts and more effective and efficient problem solving strategies.

By contrasts, using the traditional-based curriculum way of teaching, the mathematics teacher tells the students facts, demonstrates procedures, shows how to solve the problems, and then the students memorize the facts and practice the procedures.

The outcomes of the standards-based curriculum and traditional-based curriculum can be compared. If both curriculums include the same course learning objectives and academic rigor, then the main differences are the instructional method and learning environment. Research has shown that students who study under the standards-based curriculum programs have higher standardized test scores than traditional-based curriculum, and thus, in general better conceptual understanding. Standards-based curricula have positive influences on student's performance and motivation.

• 1-12 Summary

A thorough understanding of geometry, unlike most branches of mathematics, can not be accomplished by mere rote memorization. The solving of geometric problems often requires inductive and deductive reasoning and problem solving skills. The National Council of Teachers of Mathematics (NCTM) has published several documents with recommendations and guidelines that provide a framework and set of goals for mathematics curriculum improvement. The NCTM published *Principles and Standards for School Mathematics* in 2000 with specific geometry standards for grades 9-12. They also published reasoning and proof standards for grades 9-12.

A widely used model for geometry instruction and student learning was developed by a husband-and-wife team of Dutch educators, Pierre van Hiele and Dina van Hiele-Geldof (1959). These observations led them to develop a theory involving levels of thinking in geometry that students pass through as they progress from merely recognizing a figure to being able to write a formal geometric proof. As part of their model, they developed tables of levels of geometric understanding, progress indicators, and learning phases.

The Connected Mathematics Project (CPM) reform-based instructional model, "launch, explore, and summarize", is a problem-centered teaching approach based on the NCTM standards that opens the classroom to exploring, conjecturing, reasoning, and communicating. Research has shown that students who study under the standards-based curriculum programs have higher standardized test scores than traditional-based curriculum, and thus, in general better conceptual understanding. Standards-based curricula have positive influences on student's performance and motivation.

• References – Section 2

National Council of Teachers of Mathematics. (1989). *The Curriculum and Evaluation Standards*. Reston, VA: Author.
National Council of Teachers of Mathematics. (2000). *Principles and Standards for Teaching School Mathematics*. Reston, VA: Author.

Van Hiele, P. M. (1986). *Structure and Insight*. Orlando, FL: Academic Press.

Van Hiele, P. M. (1959). *The Child's Thought and Geometry*. In T. P. Carpenter, J. A. Dossey, & J. L. Koehler (Eds.), *Classics in Mathematics Education Research* (pp. 60-66). Reston, VA: National Council of Teachers of Mathematics.

Genz, Rebekah, (2006). *Determining High School Geometry Student's Geometric Understanding Using Van Hiele Levels*. Retrieved May 25, 2011, from http://contentdm.lib.byu.edu/ETD/image/etd1373.pdf

Burger, W. F., & Shaughnessy, J. M. (1986). *Characterizing the Van Hiele levels of Development in Geometry. Journal for Research in Mathematics Education, 17,* 31-48.

Halat, Jakubowski, Aydin (2008). *Reform-Based Curricum and Motivation in Geometry. Eurasia Journal of Mathematics, Science & Technology Education*, 4(3), 285-292. Retrieved May 25, 2011, from http://sabanciuniv.academia.edu/GurolIrzik/Papers/360718/Philosophy_Science_Education_and_Culture

Grading Scale: One point for each correct answer.

Excellent = 70-77, Good = 62-69, Average = 54-61, Fair = 47-53, Poor = 0-46

• 1-2 The Geometry Universe

Match definitions and terms.

A = Plane Geometry   B = Solid Geometry   C = Spherical Geometry

1. The shortest distance between points is a curved line.   _____
2. Three-dimensional objects.   _____
3. Zero, one, and two-dimensional objects on a flat surface.   _____
4. X-axis, Y-axis, and Z-axis are used.   _____
5. The plane is the universe of all points.   _____
6. Triangles with three right angles are possible.   _____
7. Two-dimensional objects on a curved surface.   _____
8. The shortest distance between points is a straight line.   _____
9. The universe is all points anywhere.   _____

• 1-3 Definitions

Match definitions and terms.

A = Point   B = Line   C = Line Segment   D = Ray   E = Angle

F = Plane   G = Width   H = Height   I = Depth   J = Dimensions

K = Polygon   L = Polyhedron

1. Volume distance of an object on an intersecting perpendicular plane.   _____
2. Space between two rays or line segments that meet at a common end point.   _____
3. Part of a line, with points marking the beginning and ending locations.   _____
4. Three-dimensional object bounded by polygons.   _____
5. Location without a size, width, height, depth, area, or volume.   _____
6. Horizontal distance on a plane.   _____
7. Two-dimensional object bounded by straight lines.   _____
8. Vertical distance on a plane.   _____

9. Line with points marking beginning and extends infinitely in one direction. ____
10. Flat surface extending infinitely in all directions. ____
11. Minimum number of coordinates needed to specify each point within an object. ____
12. Line with points marking beginning and ending locations. ____

Determine the two categories for each group of objects. Which object does not belong in each of the groups?

A = Two Dimensions   B = Three Dimensions   C = Polytopes   D = Non-polytopes

1. Triangle Circle Square Pentagon   Categories:____ ____   Object:_____
2. Sphere Cylinder Cube Cone   Categories:____ ____   Object:_____
3. Square Tetrahedron Cube Octahedron   Categories:____ ____   Object:_____
4. Circle Ellipse Ring Sphere   Categories:____ ____   Object:_____

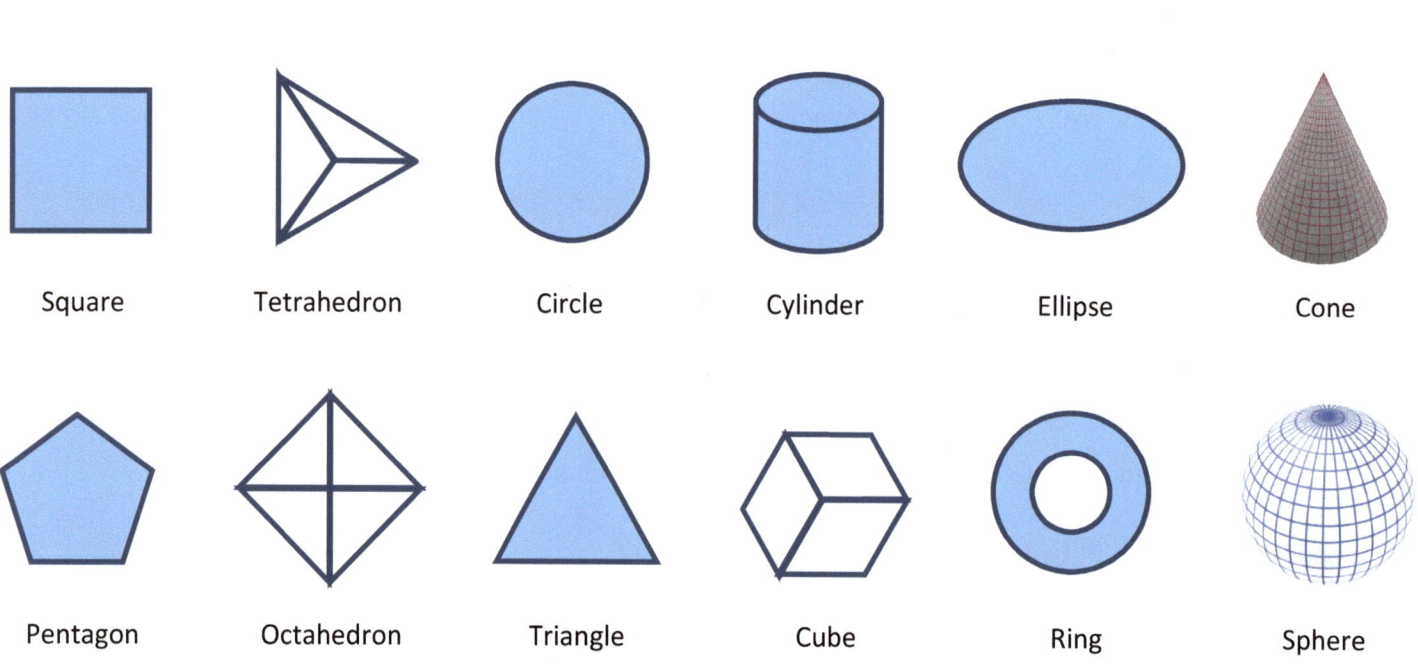

| Square | Tetrahedron | Circle | Cylinder | Ellipse | Cone |

| Pentagon | Octahedron | Triangle | Cube | Ring | Sphere |

Measure the perimeter of each object. Which object does not belong in each of the groups? What perimeter rule is used by the group in each row?

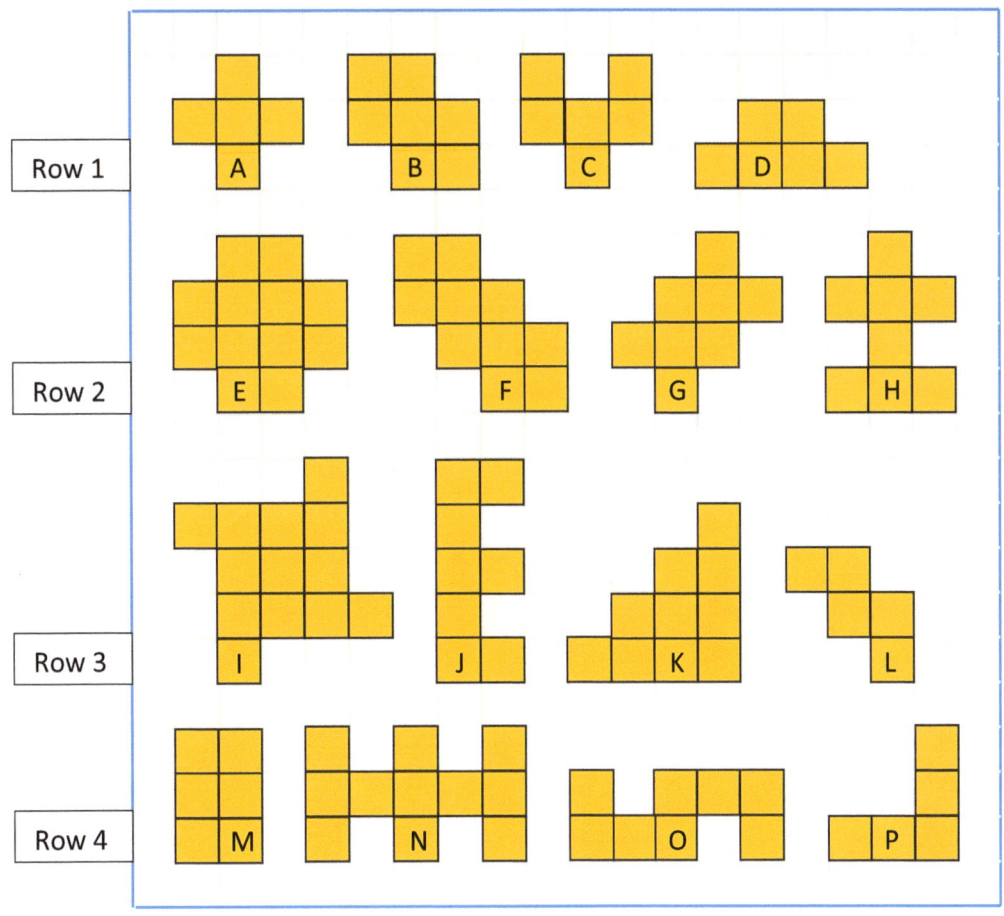

Row 1   A   B   C   D

Row 2   E   F   G   H

Row 3   I   J   K   L

Row 4   M   N   O   P

1. Perimeters: A = ____   B = ____   C = ____   D = ____

Incorrect Object: ____   Rule: _____

2. Perimeters: E = ____   F = ____   G = ____   H = ____

Incorrect Object: ____   Rule: _____

3. Perimeters: I = ____   J = ____   K = ____   L = ____

Incorrect Object: ____   Rule: _____

4. Perimeters: M = ____   N = ____   O = ____   P = ____

Incorrect Object: ____   Rule: _____

• Section 2 – Geometry Standards

• 1-8 Van Hiele Levels of Geometric Understanding

Match definitions and terms.

A = Visualization   B = Analysis   C = Abstraction
D = Deduction   E = Rigor

1. Students see figures as collections of properties.   ____
2. Students can construct proofs and understand the role of axioms and definitions.   ____
3. Students recognize figures by appearance alone.   ____
4. Students understand the formal aspects of deduction.   ____
5. Students perceive relationships between properties and between figures.   ____

• 1-9 Progress Indicators

Match definitions and terms.

A = Level 1   B = Level 2   C = Level 3   D = Level 4

1. Formation of complete definitions of types of shapes.   ____
2. Imprecise properties to compare drawings and to identify shapes.   ____
3. Clarifications of questions and rephrasing of tasks into precise language.   ____
4. Comparing shapes explicitly by means of properties of their components.   ____

• 1-10 Learning Phases

Match definitions and terms.

A = Information or inquiry   B = Guides or directed orientation
C = Explicitation   D = Free orientation   E = Integration

1. Students get acquainted with the material and begin to discover its structure.   ____
2. Students summarize what they have learned and commit it to memory.   ____

3. Students express what they have discovered and vocabulary is introduced. _____
4. Students do tasks that enable them to explore implicit relationships. _____
5. Students do more complex tasks enabling them to master relationships. _____

Match definitions and terms.

A = Launch phase   B = Explore phase   C = Summarize phase

1. Students work individually, in small groups, or occasionally as a whole class. _____
2. Begins when most students make progress towards a solution to the problem. _____
3. Teacher introduces the problem to the class. _____
4. Students discuss their solutions and share the strategies used to reach a solution. _____
5. Teacher makes sure the students understand the problem and are engaged in it. _____
6. Students work by searching for patterns, gathering data, and exchanging ideas. _____

• 2-1 Introduction

In plane geometry, an ==angle== is defined as a figure formed by two rays drawn from the same point. The common endpoint is called the vertex of the angle. The two rays are called the sides of the angle. The word *angle* comes from the Latin word *angulus*, meaning "a corner". Angles are denoted by the $\angle$ symbol. In geometry, an angle is formed by the rotation of a line about a point. The ==magnitude of the angle== is the amount of rotation that separates the two rays, and can be measured by considering the length of circular arc swept out when one ray is rotated about the vertex to coincide with the other.

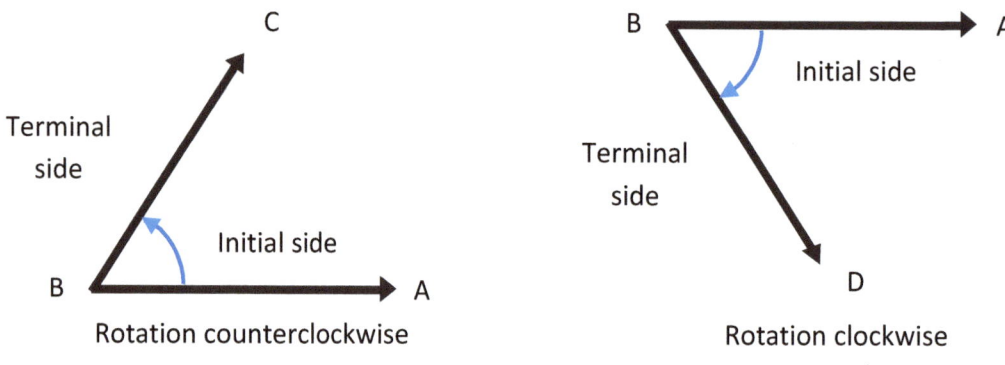

Figure 2-1: Positive Angle   Figure 2-2: Negative Angle

In the first figure above, line BA has been rotated counterclockwise into the position BC. BA is the ==initial side== of the angle and BC is the ==terminal side==. When rotation is counterclockwise, the angle formed is a positive angle. In the second figure above, the initial side BA has been rotated in a clockwise direction into position BD. The clockwise rotation forms a negative angle. In each figure, the curved arrow indicated the direction of rotation.

When two lines meet at a common point, the ==vertex==, the angle between them is called the included angle. The lines define the angle. In the above example, angle $\angle$ABC is referred to as the "included angle of BA and BC" or "BA and BC and the included angle". The angle sides are $\overrightarrow{BA}$ and $\overrightarrow{BC}$ and B is the vertex. "The angle ABC" refers to the actual angle object. If the size or measure of the angle is referenced, then "the measure of the angle ABC", written as $m\angle$ABC, should be used.

In geometry, as well as in other branches of mathematics, it is frequently necessary to describe the position of a point in a plane. This can be done by the use of a rectangular coordinate system. To use such a system, two mutually perpendicular lines are drawn. These lines, called ==coordinate axis==, intersect at a point called the ==origin==. The plane is divided into four regions called ==quadrants==. It is now possible to descibe the position of a point with respect to these axis. For example, it can be stated that a point lines 2 units above the horizontal axis and 3 units to the right of the vertical axis. The horizontal axis is called the ==x-axis== and the vertical axis is called the ==y-axis==.

An angle is in the ==standard position== when its vertex coincides with the origin of the rectangular coordinate system and its initial side lies along the positive x-axis. When the terminal side of an angle is in the first quadrant, the angle is referred to as a first quadrant angle. When the terminal side is in the second quadrant, the angle is a second quadrant angle. The same pattern holds true for the third and fourth quadrants. In the first figure below, a rectangular coordinate system is shown with a first quadrant angle.

The ==angle of rotation==, starting from the standard position, is positive upward (counterclockwise) and negative downward (clockwise). In the second figure below, the angle of rotation is shown for a complete circle about the origin point. The angle measurement is in degrees.

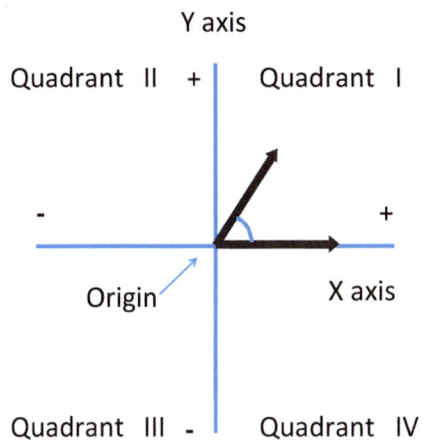

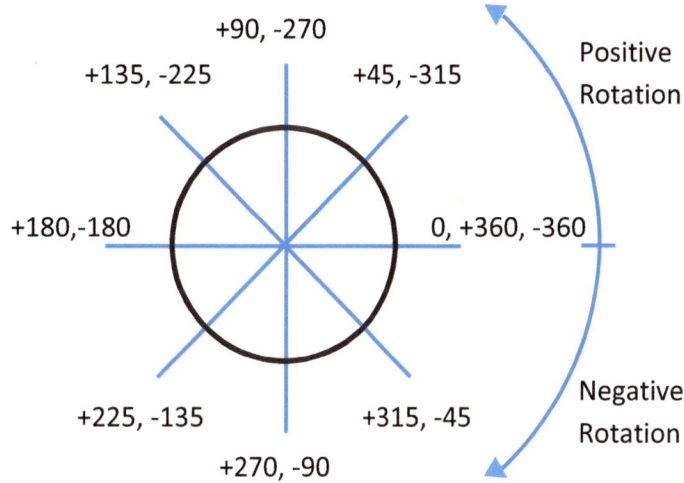

Figure 2-3: Coordinate System　　　　Figure 2-4: Angle Measurement – Degrees

2-3 Labeling Angles

An angle can be named in one of three ways:

1. Three Letters – The center letter corresponds to the vertex of the angle and the other letters representing points on the sides of the angle. In the first figure below, the name of the angle whose vertex is B can be angle ABC or angle CBA.
2. One Number – Placing a number at the vertex and in the interior of the angle. In the second figure below, the name of the angle whose vertex is B can be angle 1 or angle ABC or angle CBA.
3. One Letter – Using a single letter that corresponds to the vertex, provided that this does not cause any confusion. In the third figure below, the name of the angle whose vertex is B can be angle B. There is no question which angle on the diagram corresponds to angle B, but which angle on the diagram is angle A? Three angles are formed at vertex A: angle BAD, angle BAC, and angle CAD. To uniquely identify the angle having A as its vertex, the angle must be named using three letters or a number must be added to the diagram.

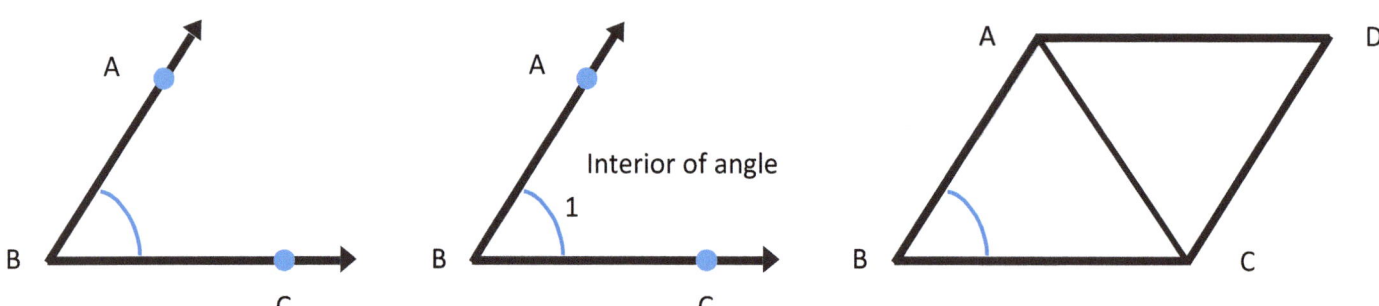

Figure 2-5: Angle ABC or Angle CBA   Figure 2-6: Angle 1   Figure 2-7: Angle B

2-4 Equivalence and Congruence

There are many mathematical relationship symbols that express comparisons between numbers and objects. The relationships may be equivalency, such as equal, not equal, greater than, less than, or other various combinations. Relationships can also include ratios and similarities. Numbers that have the same value are equal (=). Numbers that are almost equal are approximately equal (≈). Geometric objects, such as angles and polygons, which have the same shape and size, are congruent (≅). Objects which have the same shape but not the same size are similar (~). Equivalence is used to compare numbers, and congruence is used to compare objects. The figures below illustrate these four relationships.

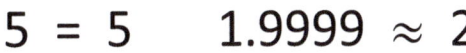 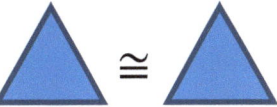 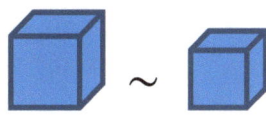

| Figure 2-8: Equal | Figure 2-9: Approximately Equal | Figure 2-10: Congruent | Figure 2-11: Similar |

Two angles are congruent if their angles are equal in measure. In geometry, two figures are congruent if they have the same shape and size. More formally, two sets of points are called congruent if, and only if, one can be transformed into the other by an ==isometry==, such as a combination of rotations, reflections, and translations. If one shape can become another by ==rotations== (turning), ==reflections== (flipping), or ==translations== (sliding), then the shapes are congruent. The figure below illustrates transformations of a triangle.

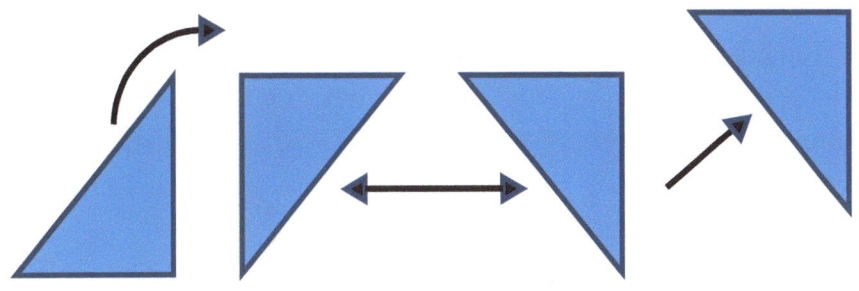

| Figure 2-12: Rotation | Figure 2-13: Reflection | Figure 2-14: Translation |

Congruent sides or segments have the exact same length. Congruent angles have the exact same measure. For any set of congruent geometric figures, corresponding sides, angles, and faces are congruent. Congruent segments, sides, and angles are often marked. In the first figure below, the four sides of the square are marked to show they are congruent to each other. In the second figure below, the two marked angles of the triangle are congruent to each other.

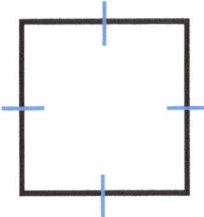 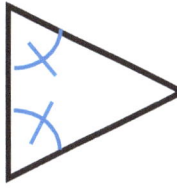

| Figure 2-15: Congruent Line Segments | Figure 2-16: Congruent Angles |

The word *congruent* comes from the Latin word *congruere*, meaning "agree, correspond with". It can be defined as equal in size and shape. Two objects are congruent if they have the same dimensions and shape. Two line segments are congruent if they have the same length. They do not need to lie at the same angle or position on the plane. Two angles are congruent if they have the same measure. The angles may lie in different orientations or positions. Two circles are congruent if they have the same size. The size can be measured as the radius, diameter, or circumference. Congruent polygons have an equal number of sides, and all of the corresponding sides and angles are congruent. They can be in different locations and transformed, and still remain congruent.

• 2-5 Angle Measurement

As mentioned earlier in this chapter, an angle is formed by the rotation of a line about a point. The amount of rotation can be measured by considering the length of the circular arc when one ray is rotated about the vertex. The number of full turns or circles around the vertex can be counted. If partial turns are made, estimates of fractional amounts or percentages can be made. The angle may be one half or one quarter of a turn, or 75% or 15% of a turn. Often an unmeasured angle is used for illustrative purposes. However, in most cases exact measurements of angles are needed.

The concept of angle is one of the most important concepts in geometry. The concepts of equality, sums, and differences of angles are important and used throughout geometry. The subject of trigonometry is based on the measurement of angles. There are two commonly used units of measurement for angles. The sexagesimal system measures angles in degrees. The circular system measures angles in radians.

• 2-6 Degrees

The most common measurement of angles is in degrees. "Sexagesimal" is a term which originated from the Sumerians in the 3rd millennium BCE, referring to a system of numerals using the number 60 as its base. The word sexagesimal is derived from the Latin word *sexagesimus,* meaning sixty. This numeral system was passed to the Babylonians, and it is still used in a modified form for measuring time, angles, and the geographic coordinates that are angles.

The reason 60 was chosen as the base number is due to the fact that it has 12 factors (1, 2, 3, 4, 5, 6, 10, 12, 15, 20, 30, 60), making it very easily divisible by many numbers, including every number from 1 through 6. For example, one hour can be divided evenly into sections of 30 minutes, 20 minutes, 15 minutes, 12 minutes, 10 minutes, 6 minutes, 5 minutes, etc. Similarly, the practical unit of measure is the ==degree==, of which there are 360 in a circle. There are 60 minutes of arc in a degree, and 60 seconds of arc in a minute.

The symbol for a degree is a small superscript circle (°). A minute is denoted with a single prime (′) and a second is denoted with a double prime (″). A ==minute== can be referred to as a minute of arc (MOA), arc minute, or just a minute. A ==second== can be referred to as a second of arc (SOA), arc second, or just a second. One advantage of the sexagesimal system is that many angles common in simple geometry are measured in whole number of degrees. Fractions of a degree can be expressed using the sexagesimal subunits of the "degree-minute-second" system or normal decimal notation. For example, 45° 52′ 30″ is equal to 45 + 52/60 + 30/3600 degrees, or 45.875 degrees. A right angle, one fourth of a turn, is 90°.

Degrees are commonly measured using a protractor. A ==protractor== is a flat curved ruler with markings that correspond to distances, usually measured in degrees or radians. Below are circular protractors for measuring angles in percentages or turns and in degrees.

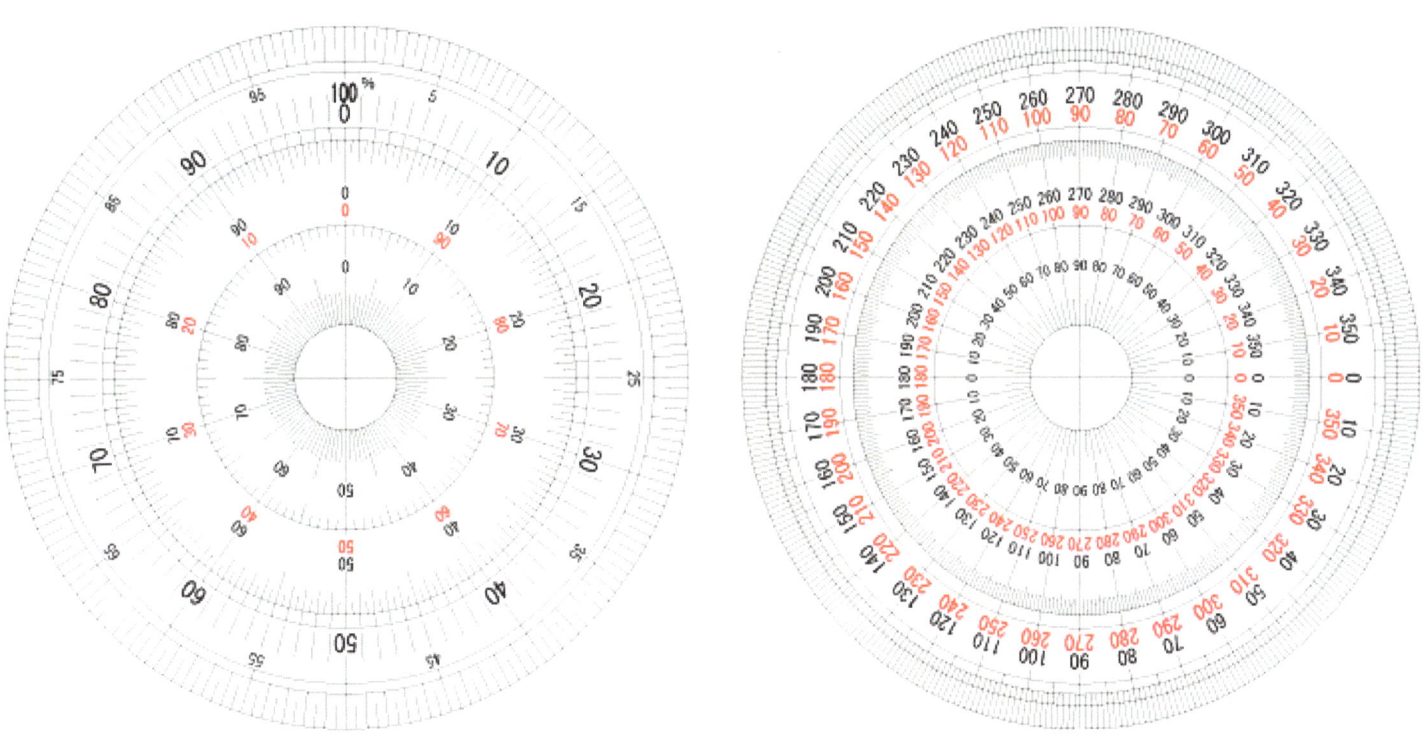

Figure 2-17:
Circular Protractor – Percentages

Figure 2-18:
Circular Protractor – Degrees

Most commonly used angles are less than one half of a turn in measurement. As a result, semi-circular protractors are usually used for angle measurement. Below are semi-circular protractors for measuring angles in degrees and radians.

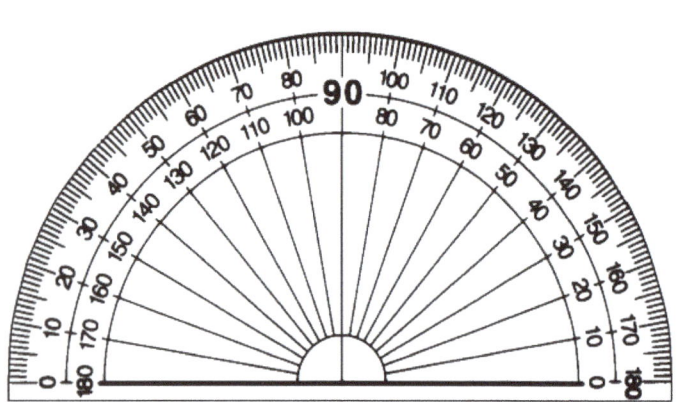

Figure 2-19:

Semi-circular Protractor – Degrees

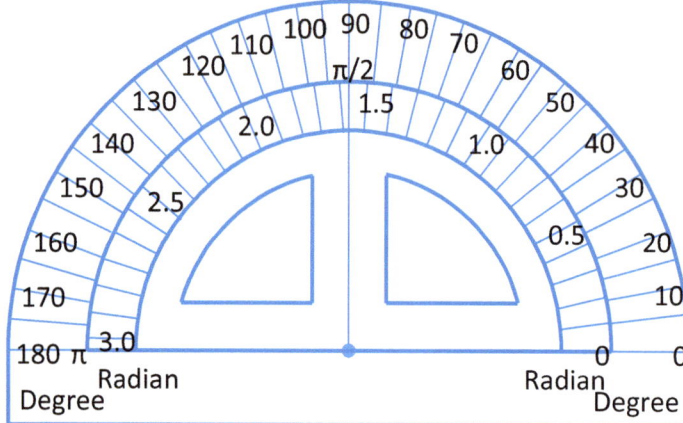

Figure 2-20:

Semi-circular Protractor – Degrees and Radians

● 2-7 Radians

A radian is the measure of an angle created by an arc that has the same length as the circle's radius. The angle in radians can also be thought of as the ratio between the arc length and the radius. If the length of the arc is 5 and the radius of the circle is 4, then the radian measure is 1.25. The arc length can be calculated by multiplying the radian measure by the radius of the circle. If the radian measure is 2 and the radius of the circle is 3, then the arc length is 6.

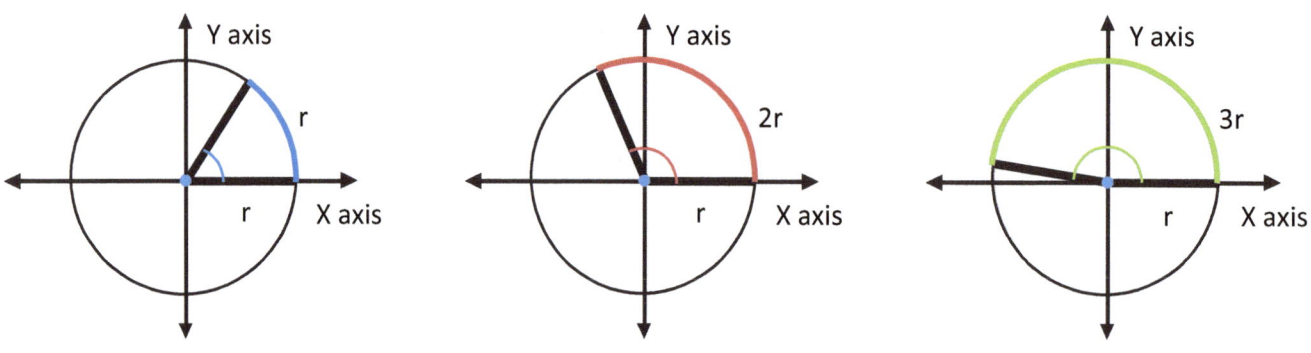

Figure 2-21: 1 Radian   Figure 2-22: 2 Radians   Figure 2-23: 3 Radians

The figures above show the central angles of three circles measured as 1 radian, 2 radians, and 3 radians. The measure of a ==central angle== is determined by the number of times the radius of the circle is contained in the intercepted arc. The measure of the angle in radians equals the length of the intercepted arc divided by the radius of the circle. A complete circle is 2π radians.

Degree measure of angles is not a "natural measure" as it was invented out of an arbitrary choice of units. In some applications, it is easier to associate the measure of an angle with the length of an arc of a circle centered at the vertex of the angle. This is the idea of radian measure, and one radian is defined as the measure of an angle in standard position whose terminal side intercepts an arc of length r units in a circle of radius r with its center at the origin. If the radius of the measuring circle is taken as 1 unit, then the length of the intercepted arc is the radian measure of the angle.

Because the circumference of a circle of radius r is $2\pi r$, there are 2π radians in a full circle. Therefore, $360° = 2\pi$ radians, and $180° = \pi$ radians. There is no standard symbol for radian measure, but when used it is customary to just give the pure number. Thus if a problem says an angle measure is 2, then the meaning is that the angle measures 2 radians. The radian is sometimes abbreviated rad. 1 radian equals approximately 57.2958 degrees.

To covert between degrees and radians, the following two formulas can be used.

Degrees = (180 * Radians) / π Radians = (π * Degrees) / 180

Using these formulas, 45 degrees is approximately equal to 0.7953 radians and 2.8 radians is approximately equal to 160.4283 degrees. The following tables show commonly used angle measurements and their equivalent values in degrees and radians.

Conversion	Exact Value	Approximate Value
Degrees to Radians	$R = \pi D / 180$	0.0174 D
Radians to Degrees	$D = 180 R / \pi$	57.2958 R

Table 2-1: Angle Conversion Formulas

Degrees	Exact Value	Approximate Value
30	π / 6	0.5235 R
45	π / 4	0.7853 R
60	π / 3	1.0471 R
90	π / 2	1.5707 R
120	2π / 3	2.0943 R
135	3π / 4	2.3561 R
150	5π / 6	2.6179 R
180	π	3.1415 R
270	3 π / 2	4.7123 R
360	2 π	6.2831 R

Table 2-2: Conversion – Degrees to Radians

Radians	Exact Value	Approximate Value
0.5	90 / π	28.6479 D
1	180 / π	57.2958 D
1.5	270 / π	85.9437 D
2	360 / π	114.5916 D
3	540 / π	171.8874 D
π	180 π / π	180.0000 D
4	720 / π	229.1833 D
5	900 / π	286.4791 D
6	1080 / π	343.7749 D
2 π	360 π / π	360.0000 D

Table 2-3: Conversion – Radians to Degrees

• 2-8 Angle Types

Angles are categorized by their size, measured in turns of a full circle, degrees, or radians. Earlier in this chapter it was stated that an angle is in the standard position when its vertex coincides with the origin of the rectangular coordinate system and its initial side lies along the positive x axis. If the angle of rotation, starting from the standard position, is zero, then the initial side and the terminal sides of the angle coincide. This is called a zero angle because the measure of the angle is 0 turns, 0 degrees, and 0 radians. An acute angle has a positive rotation of greater than 0 turns and less than 1/4 turn. The measurement in degrees is greater than 0 and less than

90, and the measurement in radians is greater than 0 and less than π/2. A right angle is 1/4 turn, 90 degrees, and π/2 radians. The ⌐ symbol is used to show a right angle. An obtuse angle is greater than 1/4 turn and less than 1/2 turn. A straight angle is 1/2 turn. A reflex angle is greater than 1/2 turn and less than 1 turn. A full angle is 1 turn, 360 degrees, and 2π radians.

In the table below, the names of the angles types and their measurement sizes is shown.

Name	Zero	Acute	Right	Obtuse	Straight	Reflex	Full
Degrees	X=0	0 <X<90	X=90	90<X<180	X=180	180<X<360	X=360
Radians	X=0	0<X<π/2	X=π/2	π/2<X<π	X=π	π<X<2π	X=2π
Turns	X=0	0<X<1/4	X=1/4	1/4<X<1/2	X=1/2	1/2<X<1	X=1

Table 2-4: Angle Types and Measurements

In the figures below, the angle of rotation is shown for each angle type.

Figure 2-24:	Figure 2-25:	Figure 2-26:	Figure 2-27:	Figure 2-28:	Figure 2-29:	Figure 2-30:
Zero Angle	Acute Angle	Right Angle	Obtuse Angle	Straight Angle	Reflex Angle	Full Angle

The zero angle, right angle, straight angle, and full angle define specific angle measurements. The acute angle, obtuse angle, and reflex angle are ranges of values for angle measurements. For example, 15 degrees, 30 degrees, 45 degrees, and 60 degrees are all acute angles. A right angle is always 90 degrees. A full angle is always 360 degrees. An angle that is not a right angle, or a multiple of a right angle, is called an oblique angle. All acute and obtuse angles are oblique angles. All reflex angles, except those equal to 3/4 turn, are oblique angles.

Angles that have the same measure or the same magnitude are congruent. In the first figure below, the measure of angle BAC is congruent to the measure of angle EDF. Both angles are 45 degrees. In the second figure below, $\vec{AC}$ is an <mark>angle bisector</mark> which divides angle BAD into two equal angles. Angle BAD is 60 degrees, angle BAC is 30 degrees, and angle CAD is 30 degrees.

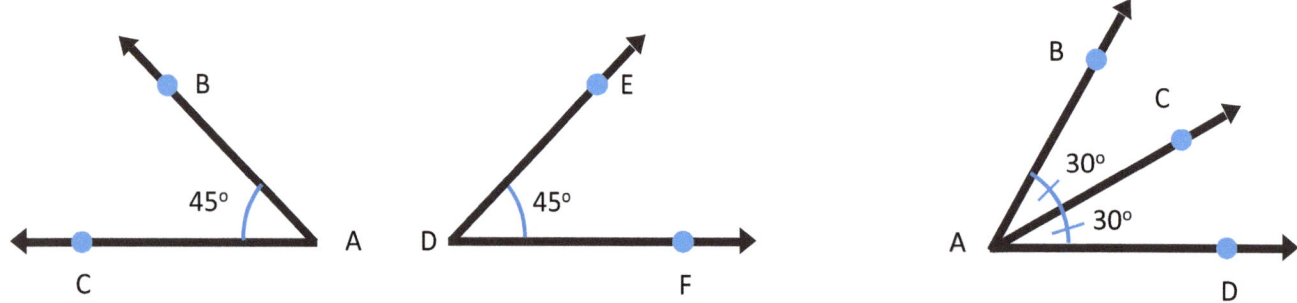

Figure 2-31: Congruent Angles   Figure 2-32: Angle Bisector

<mark>Adjacent angles</mark> share a common vertex and edge but do not share any interior points. In the first figure below, angle BAC and angle CAD are adjacent angles. <mark>Vertical angles</mark> are formed when two lines intersect and form four angles. The vertically opposite angles are equal in measure. In the second figure below, angle EAB and angle DAC are vertical angles, and equal to each other. Angle EAD and angle BAC are also vertical angles, and equal to each other. Both pairs of vertical angles, four angles together, always sum to a full angle.

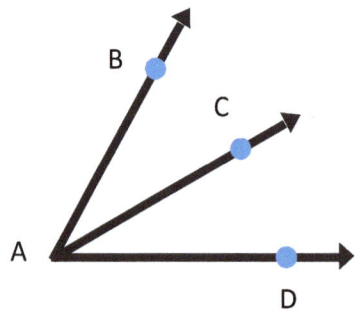

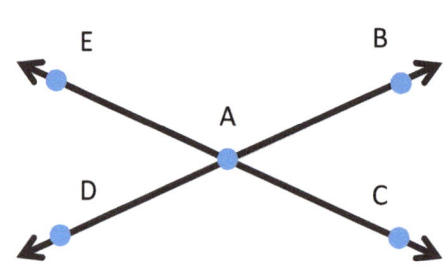

Figure 2-33: Adjacent Angles   Figure 2-34: Vertical Angles

Two angles that sum to one right angle are <mark>complementary angles</mark>. The difference between an angle and a right angle is the complement of the angle. Two angles that sum to a straight angle are <mark>supplementary angles</mark>. The difference between an angle and a straight angle is the supplement of the angle. Two angles that sum to a full angle are <mark>explementary angles</mark>. The difference between an angle and a full angle is the explement of the angle.

In the first figure below, the total measurement of the two angles is 90 degrees. In the second figure below, the total measurement of the two angles is 180 degrees. In the third figure below, the total measurement of the two angles is 360 degrees. Angles do not need to be adjacent angles to be complementary, supplementary, or explementary angles. When vertical angles are formed, each pair of adjacent angles is supplementary.

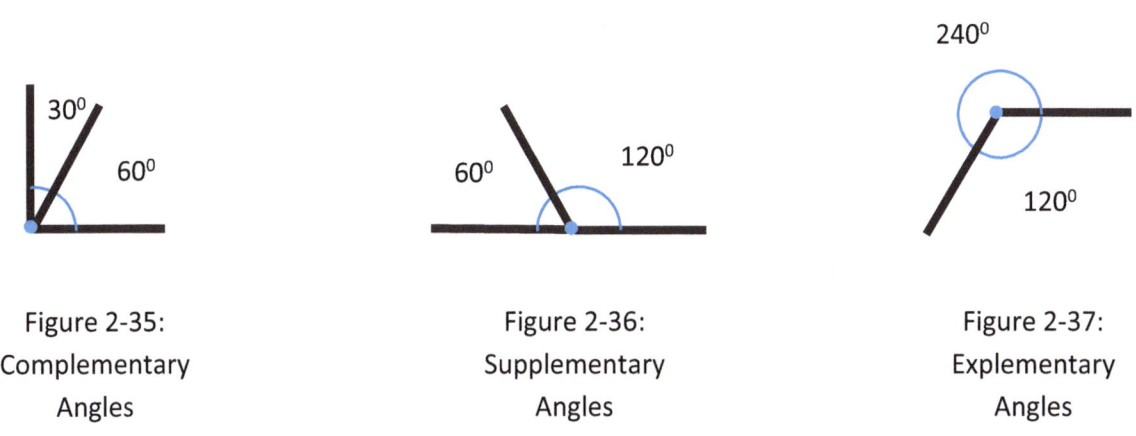

Figure 2-35:
Complementary
Angles

Figure 2-36:
Supplementary
Angles

Figure 2-37:
Explementary
Angles

Lines are often drawn and intersect to form angles. One example, as stated earlier in this chapter, is when two lines intersect to form vertical angles. Lines that are always the same distance apart and do not intersect are called <mark>parallel lines</mark>. The ‖ symbol is used for parallel lines. Lines that meet at a right angle are <mark>perpendicular lines</mark>. The ⊥ symbol is used for perpendicular lines. The lines are also referred to as being <mark>orthogonal</mark> because they form right angles. Intersecting lines that are not perpendicular to one another are <mark>oblique lines</mark>.

A line that crosses parallel lines is called a <mark>transversal line</mark>. If the transversal line is perpendicular to both of the other two lines, then the lines are parallel. If both of the angles on the same side of the transversal line are less than right angles, then the lines are not parallel. When a transversal line intersects two parallel lines, the corresponding angles at the point of intersection are equal in size. <mark>Alternate angles</mark> are also equal. In the figures below, a transversal line crosses two lines. The first figure shows parallel lines. The second figure shows nonparallel lines.

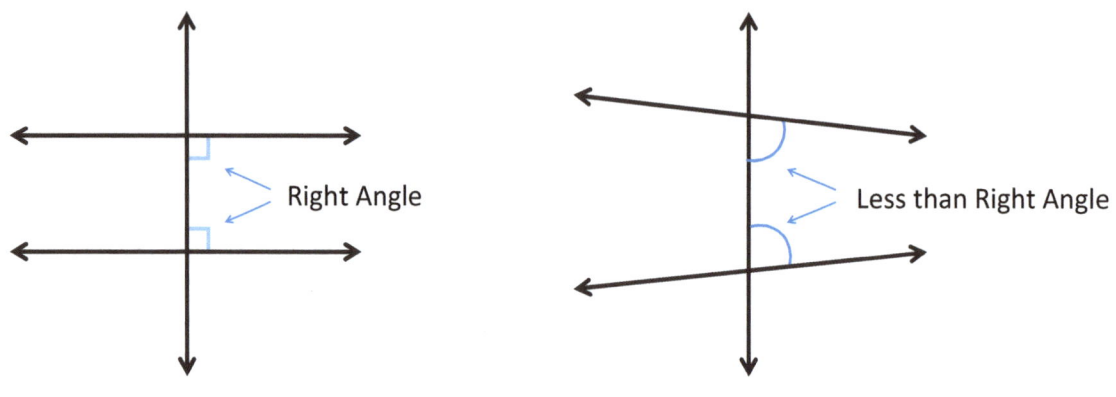

Figure 2-38: Parallel Lines   Figure 2-39: Not Parallel Lines

Interior angles are those angles formed between the two lines that are crossed by the transversal line. In the first figure below, the interior angles are angles A, B, C, and D. **Exterior angles** are those angles formed above and below the two lines that are crossed by the transversal line. In the second figure below, the exterior angles are angles E, F, G, and H.

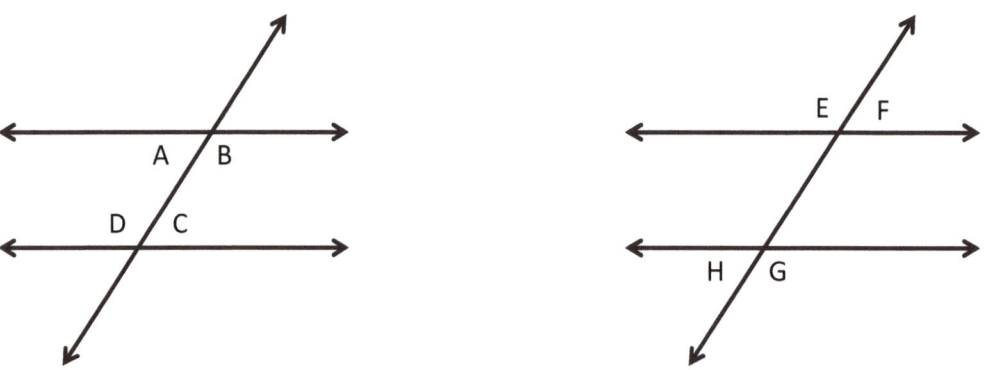

Figure 2-40: Interior Angles   Figure 2-41: Exterior Angles

The many angles created can be made into **pairs of angles** which have special names and relationships. In the first figure below, the pairs of angles are angles A through H. In the second figure below, the measurement for each angle in degrees is shown.

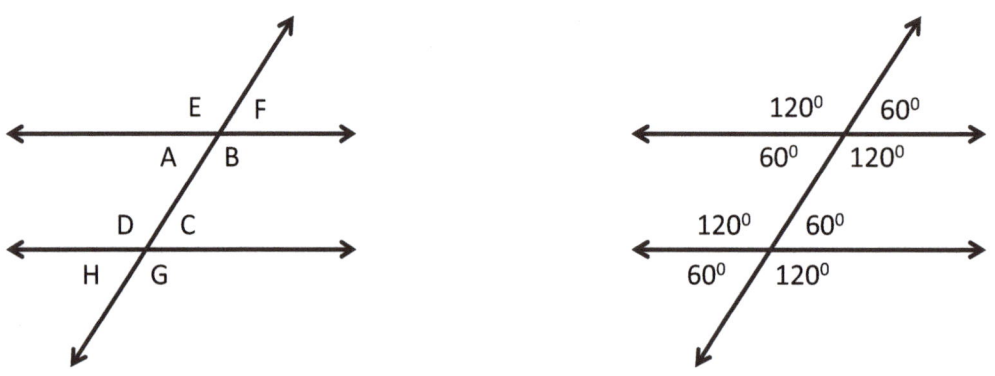

Figure 2-42: Pairs of Angles   Figure 2-43: Pairs of Angles - Degrees

The angle pairs A and C, and B and D, are called ==alternate interior angles==. Alternate interior angles have the same measurement. The angle pairs E and G, and F and H, are ==alternate exterior angles==. Alternate exterior angles have the same measurement. The angle pairs A and H, B and G, C and F, and D and E are ==corresponding angles==. Corresponding angles have the same measurement. The angle pairs A and F, B and E, C and H, and D and G are ==vertically opposite angles==. The vertically opposite angles are equal in measure. The angle pairs A and D, and B and C are ==consecutive interior angles==. The sum of consecutive interior angles is a straight angle or 180 degrees, if the two lines are parallel. The sum of any two ==adjacent angles== is a straight angle or 180 degrees. The sum of the four intersection angles, created by an ==intersection of lines==, is a full angle or 360 degree.

The table below summarizes the angle pairs by angle type.

Angle Type	Angles
Vertical - Congruent	A and F, B and E, C and H, D and G
Corresponding – Congruent	A and H, B and G, C and F, D and E
Alternate Interior – Congruent	A and C, B and D
Alternate Exterior – Congruent	E and G, F and H
Consecutive Interior – If sum equals 180^0, then parallel.	A and D, B and C
Adjacent – Sum of two equals 180^0.	F and E, E and A, A and B, B and F, C and D, D and H, H and G, G and C
Intersection – Sum of four equals 360^0.	A, B, E, and F; C, D, G, and H

Table 2-5: Angle Pairs by Angle Type

The two lines crossed by the transversal line can be tested for ==parallelism==. If any of the following cases are true, the lines are parallel: (1) any pair of alternative interior angles is equal, (2) any pair of alternative exterior angles is equal, (3) any pair of corresponding angles is equal, or (4) any pair of consecutive interior angles sums to a straight angle. If the two lines are not parallel, then only the vertical angles are equal in measure. The other corresponding angle pairs are not equal in measure.

In plane geometry, an angle is defined as a figure formed by two rays drawn from the same point. The common endpoint is called the vertex of the angle. The two rays are called the sides of the angle. In geometry, an angle is formed by the rotation of a line about a point. When two lines meet at a common point, the vertex, the angle between them is called the included angle.

In geometry, as well as in other branches of mathematics, it is frequently necessary to describe the position of a point in a plane. This can be done by the use of a rectangular coordinate system. An angle is in the standard position when its vertex coincides with the origin of the rectangular coordinate system and its initial side lies along the positive x axis. The angle of rotation, starting from the standard position, is positive upward (counterclockwise) and negative downward (clockwise).

An angle can be named in one of three ways: three letters, one number, or one letter. There are many mathematical relationship symbols that express comparisons between numbers and objects. Some important symbols used in geometry are for equal (=), approximately equal (≈), congruent (≅), and similar (~). Two angles are congruent if their angles are equal in measure. In geometry, two figures are congruent if they have the same shape and size. Congruent sides or segments have the exact same length. Congruent angles have the exact same measure. For any set of congruent geometric figures, corresponding sides, angles, and faces are congruent. Congruent segments, sides, and angles are often marked.

There are two commonly used units of measurement for angles. The sexagesimal system measures angles in degrees. The circular system measures angles in radians. The symbol for a degree is a small superscript circle (°). A second is denoted with a single prime (′) and a minute is denoted with a double prime (″). The practical unit of measure is the degree, of which there are 360 in a circle. Degrees are commonly measured using a protractor. A protractor is a flat curved ruler with markings that correspond to distances, usually measured in degrees or radians. A radian is the measure of an angle created by an arc that has the same length as the circle's radius. The angle in radians can also be thought of as the ratio between the arc length and the radius. The measure of the angle in radians equals the length of the intercepted arc divided by the radius of the circle. A complete circle is 2π radians. 1 radian equals approximately 57.2958 degrees.

Angles are categorized by their size, measured in turns of a full circle, degrees, or radians. Based on the angle of rotation, angles are described as being: zero, acute, right, obtuse, straight, reflex, or full. Angles that have the same measure or the same magnitude are congruent. Adjacent angles share a common vertex and edge but do not share any interior points. Vertical angles are formed when two lines intersect and form four angles. The vertically opposite angles are equal in measure.

Two angles that sum to one right angle are complementary angles. The difference between an angle and a right angle is the complement of the angle. Two angles that sum to a straight angle are supplementary angles. The difference between an angle and a straight angle is the supplement of the angle. Two angles that sum to a full angle are explementary angles. The difference between an angle and a full angle is the explement of the angle.

Lines that are always the same distance apart and do not intersect are called parallel lines. Lines that meet at a right angle are perpendicular lines. The lines are also referred to as being orthogonal because they form right angles. Intersecting lines that are not perpendicular to one another are oblique lines. A line that crosses parallel lines is called a transversal line. When a transversal line intersects two parallel lines, the corresponding angles at the point of intersection are equal in size. Alternate angles are also equal. Interior angles are those angles formed between the two lines that are crossed by the transversal line. Exterior angles are those angles formed above and below the two lines that are crossed by the transversal line.

The many angles created can be made into pairs of angles which have special names and relationships. Alternate interior angles have the same measurement. Alternate exterior angles have the same measurement. Corresponding angles have the same measurement. The vertically opposite angles are equal in measure. The sum of consecutive interior angles is a straight angle or 180 degrees, if the two lines are parallel. The sum of any two adjacent angles is a straight angle or 180 degrees. The sum of the four intersection angles, created by an intersection of lines, is a full angle or 360 degree. The two lines crossed by the transversal line can be tested for parallelism. If the two lines are not parallel, then only the vertical angles are equal in measure. The other corresponding angle pairs are not equal in measure.

Grading Scale: One point for each correct answer.

Excellent = 178-197, Good = 158-177, Average = 138-157, Fair = 119-137, Poor = 0-118

● 2-2 Coordinate System

Coordinates of points (X, Y)

Part 1 – Identify Quadrant

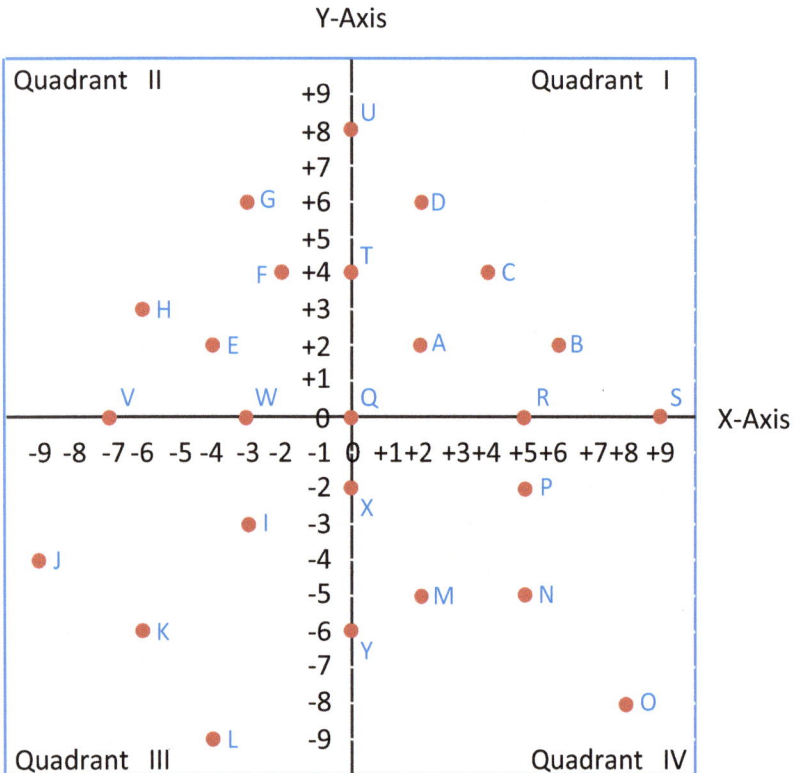

Write the quadrant or on which axis each of the 25 points is located.

1. P _____	2. O _____	3. I _____	4. U _____	5. Y _____
6. T _____	7. R _____	8. E _____	9. W _____	10. Q _____
11. A _____	12. S _____	13. D _____	14. F _____	15. G _____
16. H _____	17. J _____	18. K _____	19. L _____	20. M _____
21. N _____	22. B _____	23. V _____	24. C _____	25. X _____

Part 2 – Write Coordinates

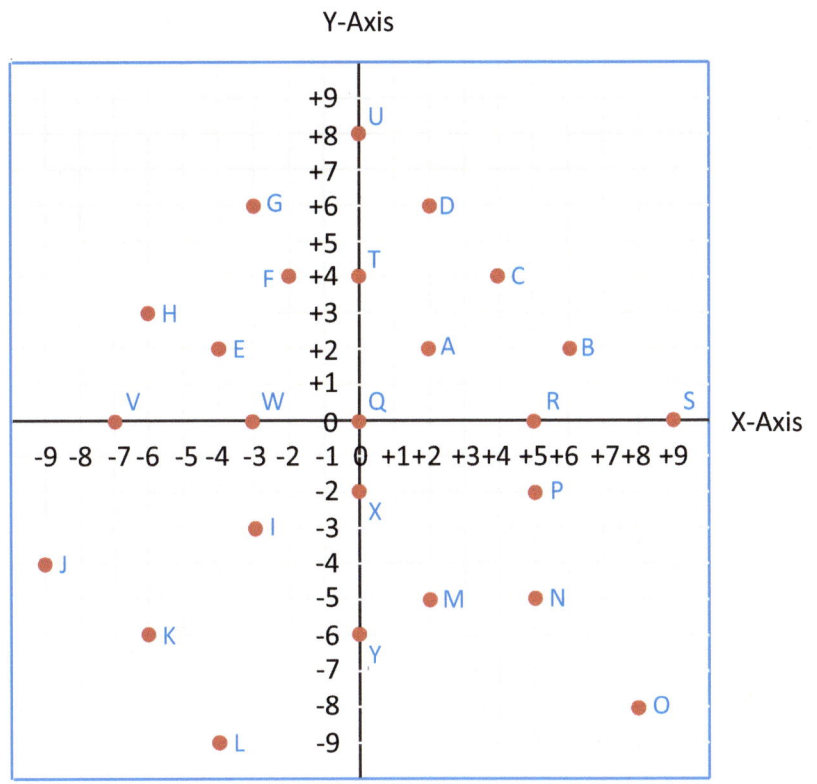

Y-Axis

X-Axis

Write the coordinates in (X, Y) format for each of the 25 points.

1. P _____   2. O _____   3. I _____   4. U _____   5. Y _____

6. T _____   7. R _____   8. E _____   9. W _____   10. Q _____

11. A _____   12. S _____   13. D _____   14. F _____   15. G _____

16. H _____   17. J _____   18. K _____   19. L _____   20. M _____

21. N _____   22. B _____   23. V _____   24. C _____   25. X _____

Part 3 – Identify Points

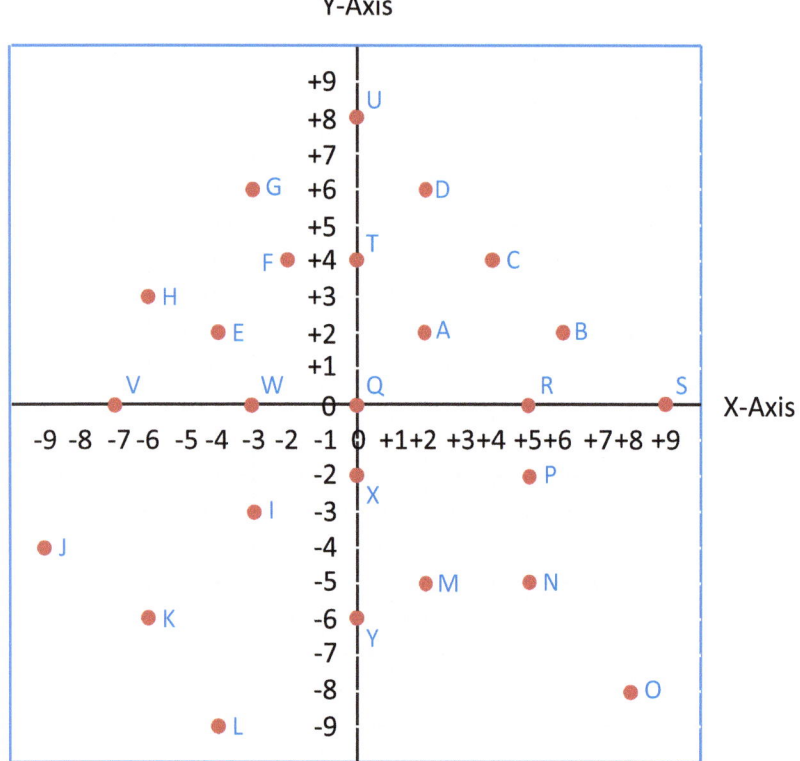

Write the letter of the point located at each of the 25 coordinates.

1. (2, 2) _____   2. (-4, 2) _____   3. (-9, -4) _____   4. (8, -8) _____   5. (0, 4) _____

6. (0, -6) _____   7. (0, 0) _____   8. (-3, 6) _____   9. (-3, 0) _____   10. (2, -5) _____

11. (4, 4) _____   12. (5, -5) _____   13. (-3, -3) _____   14. (0, -2) _____   15. (2, 6) _____

16. (9, 0) _____   17. (0, 8) _____   18. (-6, -6) _____   19. (6, 2) _____   20. (-2, 4) _____

21. (-7, 0) _____   22. (-4, -9) _____   23. (5, 0) _____   24. (5, -2) _____   25. (-6, 3) _____

Part 4 – Mark Points

Y-Axis

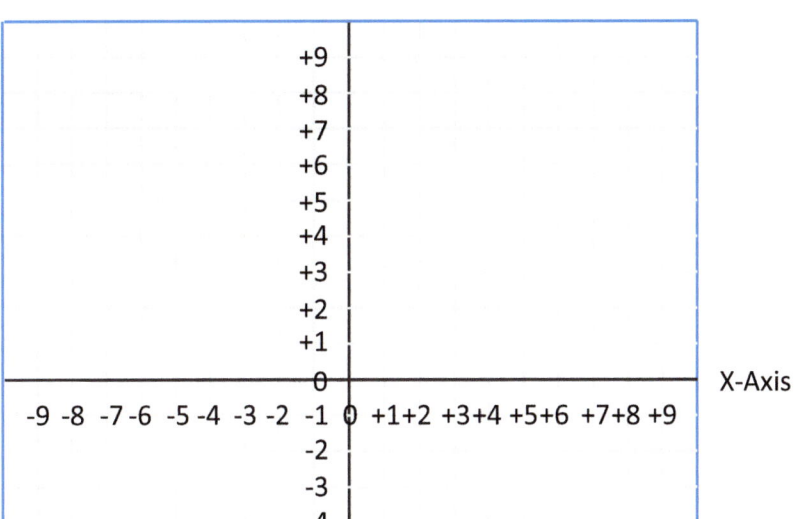

X-Axis

Mark and label each of the 25 points on the graph.

1. A (-2, 8)   2. B (-7, 0)   3. C (-5, -5)   4. D (0, -6)   5. E (7, -7)

6. F (8, 0)   7. G (9, 9)   8. H (0, 5)   9. I (-4, 5)   10. J (-3, 0)

11. K (-7, -3)   12. L (0, -2)   13. M (3, -6)   14. N (4, 0)   15. O (7, 3)

16. P (0, 1)   17. Q (-5, 4)   18. R (-2, -2)   19. S (4, -4)   20. T (1, 1)

21. U (-8, 2)   22. V (-3, -7)   23. W (6, -3)   24. X (3, 7)   25. Y (0, 0)

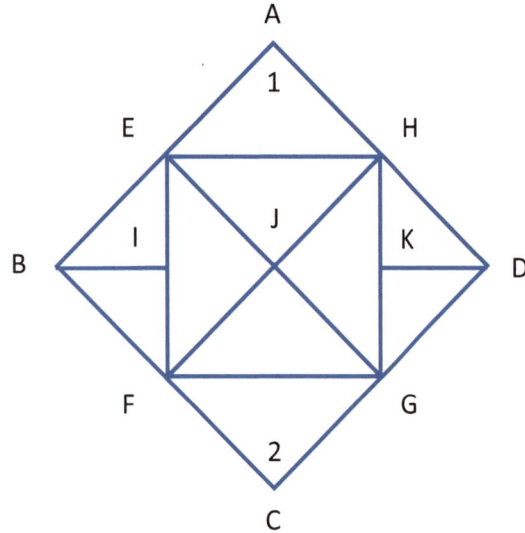

Write the angle names for each selected angle.

1. Four names for the angle with vertex of A.   _____ _____ _____ _____
2. Four names for the angle with vertex of C.   _____ _____ _____ _____
3. Three different angles with vertex B.   _____ _____ _____
4. Three different angles with vertex D.   _____ _____ _____
5. Three different angles with vertex I.   _____ _____ _____
6. Six different angles with vertex J.   _____ _____ _____ _____ _____ _____
7. Ten different angles with vertex H.

_____ _____ _____ _____ _____ _____ _____ _____ _____ _____

Translations – Draw the triangle after translation.

Part 1 – Rotation

1. Rotate the triangle +90⁰.

2. Rotate the triangle -180⁰.

3. Rotate the triangle +270⁰.

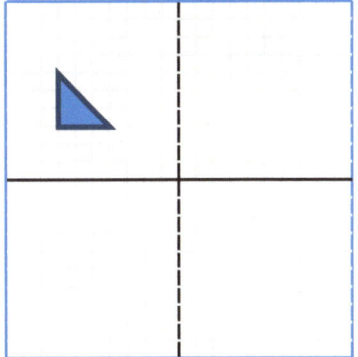 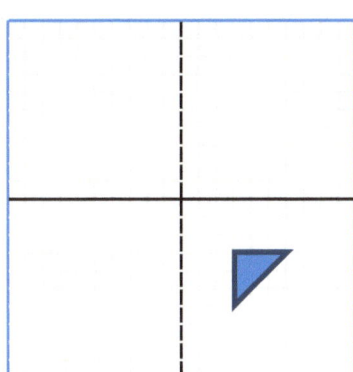 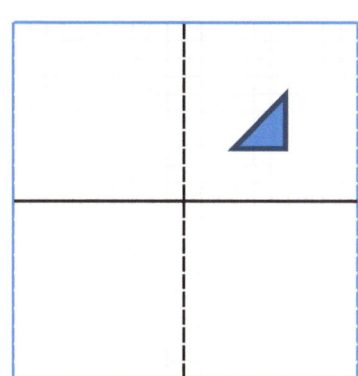

Part 2 – Reflection

4. Reflect the triangle over the x-axis.

5. Reflect the triangle over the y-axis.

6. Reflect the triangle over the x-axis.

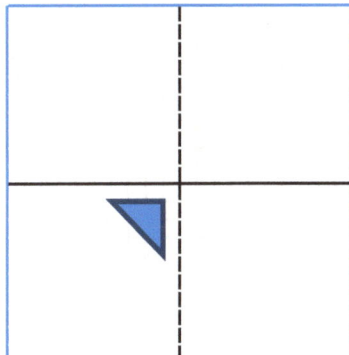 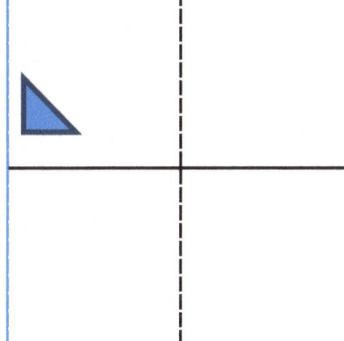 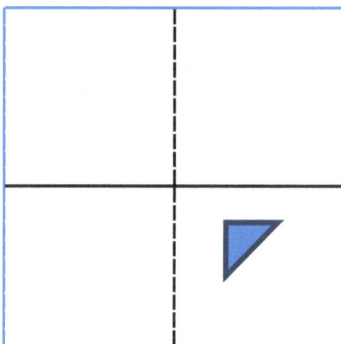

Part 3 – Translation

7. Translate the triangle left 5 units, then up 2 units.

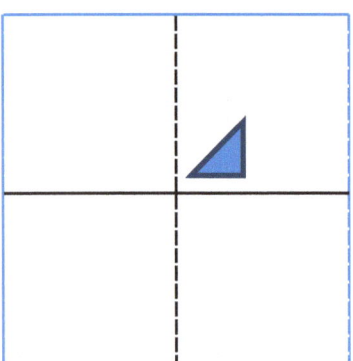

8. Translate the triangle right 3 units, then down 4 units.

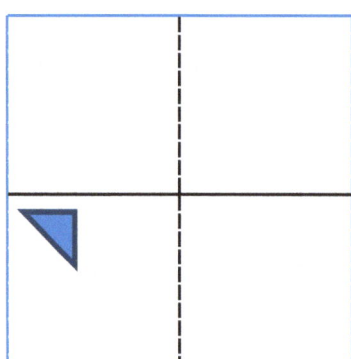

9. Translate the triangle right 6 units, then up 6 units.

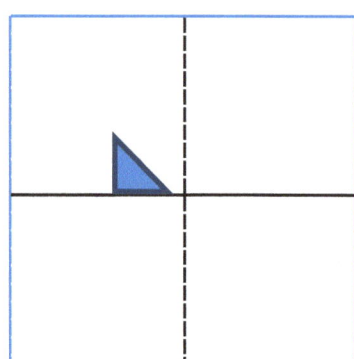

Part 4 – Combination

10. Rotate the triangle -90^0, then reflect over the x-axis.

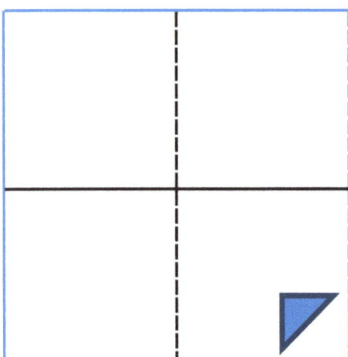

11. Reflect over the y-axis, then translate 3 left, 2 up.

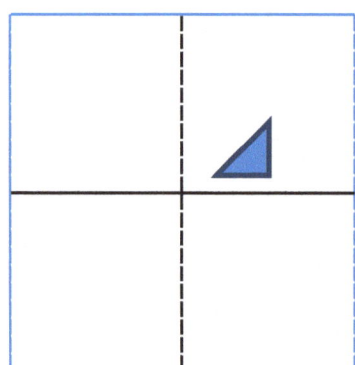

12. Translate right 2, up 4, then rotate -270^0.

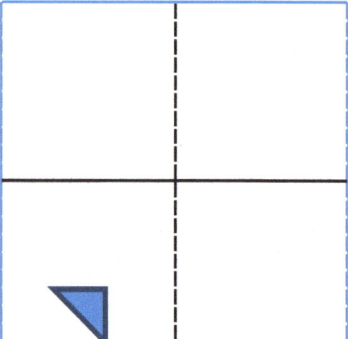

47

Use a protractor to measure the degrees of each angle.

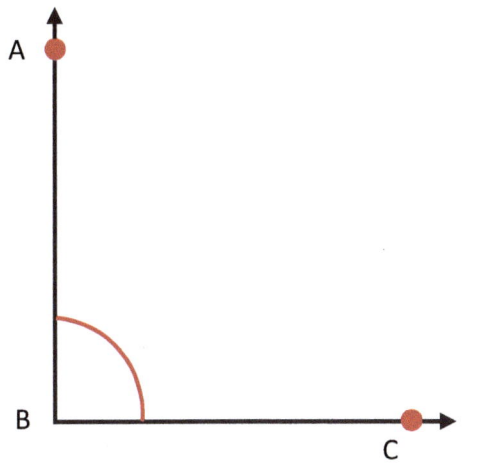

1. Angle ABC = _____ degrees

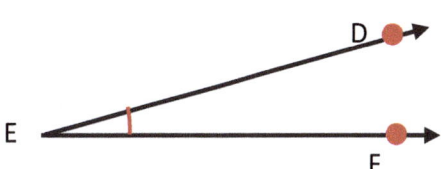

2. Angle DEF = _____ degrees

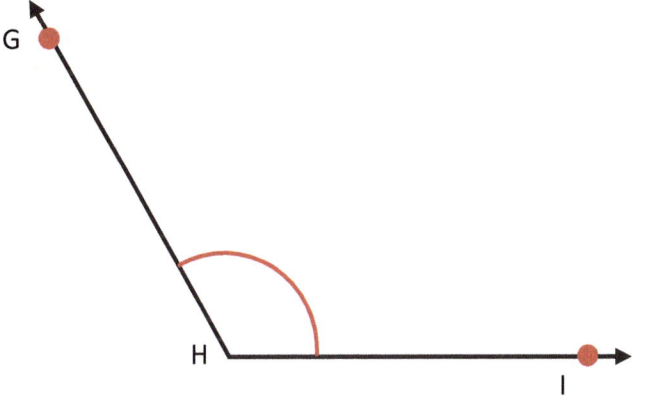

3. Angle GHI = _____ degrees

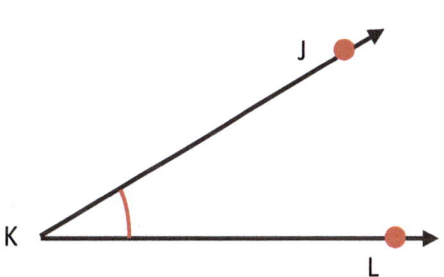

4. Angle JKL = _____ degrees

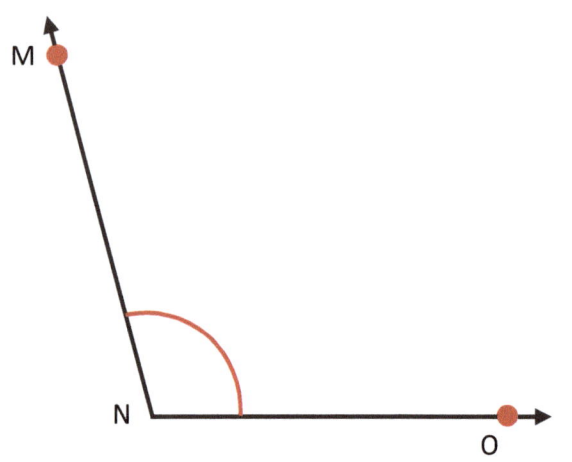

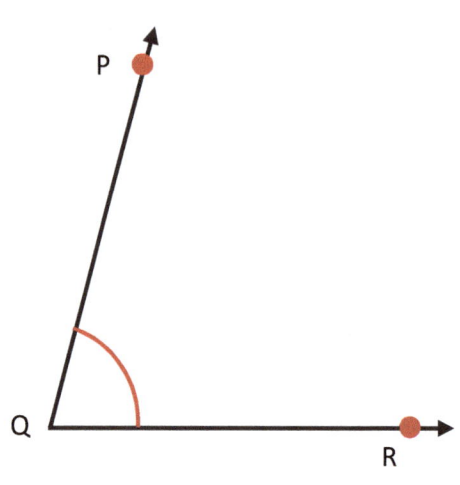

5. Angle MNO = _____ degrees

6. Angle PQR = _____ degrees

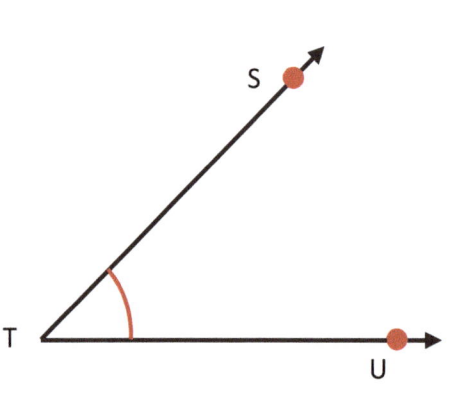

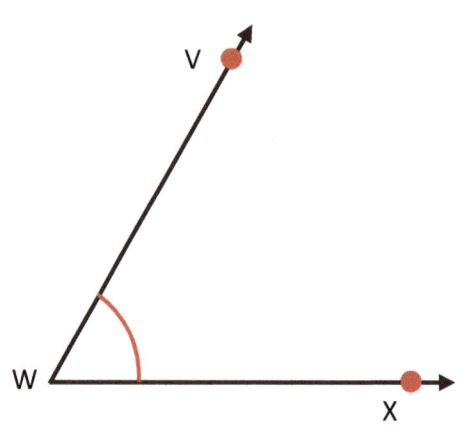

7. Angle STU = _____ degrees

8. Angle YWX = _____ degrees

2-6 Degrees

Convert angle measurements in degrees to decimal fractions of a degree.

1. 25^0 45'   _____

2. 82^0 30' 18"   _____

3. 120^0 24'   _____

4. 180^0 12' 42"   _____

5. 275^0 6'   _____

Convert angle measurements in degrees to degrees, minutes, and seconds.

6. 317.875^0 _____
7. 9.125^0 _____
8. 46.36^0 _____
9. 75.67^0 _____
10. 153.48^0 _____

2-7 Radians

Convert angle measurements in degrees to radians. (Round to the nearest thousandth.)

1. 20^0 _____ radians
2. 75^0 _____ radians
3. 160^0 _____ radians
4. 210^0 _____ radians
5. 320^0 _____ radians

Convert angle measurements in radians to degrees. (Round to the nearest thousandth.)

6. 0.75 radians _____ degrees
7. 2.5 radians _____ degrees
8. 3.25 radians _____ degrees
9. 4.5 radians _____ degrees
10. 5.75 radians _____ degrees

2-8 Angle Types

Determine the angle type as a zero, acute, right, obtuse, straight, reflex, or full angle.

1. 1/4 turns _____
2. 360 degrees _____
3. 0 radians _____
4. 3/4 turns _____
5. 45 degrees _____
6. 2π radians _____
7. 1/2 turns _____
8. 135 degrees _____
9. 1π radians _____
10. 90 degrees _____

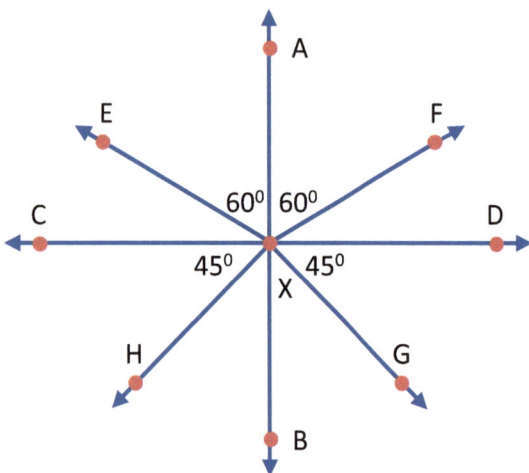

Determine the angle measurements and angle relationships. Lines AB and CD are mutually perpendicular.

1. Measurement of angle FXD, in degrees.   _____
2. Measurement of angle HXB, in degrees.   _____
3. Measurement of angle EXH, in degrees.   _____
4. Measurement of angle FXB, in degrees.   _____
5. Two angles complementary to angle CXE.   _____   _____
6. Three angles complementary to angle BXG.   _____   _____   _____
7. Two angles supplementary to angle AXE.   _____   _____
8. Three angles supplementary to angle DXB.   _____   _____   _____